CW01018052

Cardiac Output Measurement

Cardiac Output Measurement

Edited by

J. I. S. Robertson MD, FRCP (Lond.), FRCP (Glas.), FRS (Ed.).
Visiting Professor of Medicine, Prince of Wales Hospital, Chinese University of Hong Kong;
Senior Consultant in Cardiovascular Medicine, Janssen Research Foundation, Belgium; Chairman,
Working Group on Hypertension and the Heart, European Society of Cardiology.

W. H. Birkenhäger MD, PhD.
Professor of Medicine, Erasmus University, Rotterdam, The Netherlands;
President, European Society of Hypertension.

SAUNDERS

London Philadelphia Toronto Sydney Tokyo

W. B. Saunders
Baillière Tindall

24–28 Oval Road
London NW1 7DX, England

The Curtis Center
Independence Square West
Philadelphia, PA 19106–3399, USA

55 Horner Avenue
Toronto, Ontario M8Z 4X6, Canada

Harcourt Brace Jovanovich Group
(Australia) Pty Ltd
30–52 Smidmore Street
Marrickville, NSW 2204, Australia

Harcourt Brace Jovanovich Japan Inc.
Ichibancho Central Building,
22–1 Ichibancho
Chiyoda-ku, Tokyo 102, Japan

This book is printed on acid-free paper.

British Library Cataloguing in Publication Data is available from the British Library

ISBN 0-7020-1573-3

Contents

Preface

The clinical measurement of cardiac output is a frequent requirement in a wide range of circumstances, ranging from the evaluation of heart failure to the treatment of patients undergoing intensive care. Cardiac output is also a crucial, albeit elusive, component of raised arterial pressure. Despite the needs, accurate, reliable and repeated determination of cardiac output remains difficult to achieve. Most obviously, some of the invasive techniques employed in the quest for accuracy may so alarm the subject that they can induce bigger changes in cardiac output than they are intended to assess. Pickering commented in 1955 'The only completely reliable method is that involving the Fick principle and the use of cardiac catheterization, and this requires a formidable array of instruments, masked men, and punctures which are not without their effects on the circulation of the conscious man ... discussion on cardiac output illustrates with great clarity the difficulty of considering any single characteristic of the circulation except in relationship to the circulation as a whole and with the proviso, which is very rarely justified, that all other factors remain unaltered ... the frustrated investigator all too often finds that his arguments closely imitate the behaviour of the blood in following a circular course'[1].

Such astringent scepticism from so prominent a critic had an inevitably depressing effect on efforts to measure cardiac output. As late as 1990 Hainsworth could comment 'A reliable non-invasive method of estimating cardiac output has long been a "holy grail" in cardiovascular physiology, and the multiplicity of methods, involving many physical and chemical approaches, attests to the difficulties'[2].

Such caution is to be contrasted with the blithe insouciance of others, who pursue cardiac output measurement with apparent ignorance of or disregard for the problems. Since such results are not infrequently published, these workers evidently succeed on occasion in misleading referees and editors, as well as themselves.

The present volume was conceived to provide a critical contemporary review of these problems, and explicitly to define criteria for the clinical and experimental measurement of cardiac output in man. It was intended to give an appreciation of how well or badly present methods approximate to the ideals. A major objective was to set high but realistic standards, and to discourage pollution of the literature with results obtained using invalidated techniques or with lax application.

This volume of work was brought together by the Working Group on Hypertension and the Heart of the European Society of Cardiology and the European Society of Hypertension, with the support of the Janssen International Research Council. Over-enthusiastic advocacy of any particular technique was intended to be countered by having at least two, and sometimes more, contributions on each topic.

The book comprises a critical review of the major approaches currently employed for the clinical determination of cardiac output. There are additional sections on systolic time interval measurements and on diastolic function evaluation methods which, while not measuring cardiac output as such, are often employed in the clinical assessment of cardiac function. Although the principal focus is on man, work in animals is also discussed where it may serve to illuminate certain aspects. Finally, there are brief reviews of the field by two experienced critics, Drs Stevo Julius and James Conway.

We believe that this publication constitutes a unique and comprehensive contemporary analysis of this important and most difficult problem.

J. I. S. ROBERTSON — *Hong Kong & Beerse*

W. H. BIRKENHÄGER — *Rotterdam*

References

[1] Pickering GW. High blood pressure. London: Churchill 1955: 48–52
[2] Hainsworth R. Non-invasive investigations of cardiovascular reflexes in humans. Clin Sci 1990; 78: 437–43.

Measurement of cardiac output: Fick principle using catheterization

R. FAGARD* AND J. CONWAY**

*Hypertension and Cardiovascular Rehabilitation Unit, Department of Pathophysiology, Faculty of Medicine, University of Leuven, Leuven, Belgium and **Department of Cardiology, John Radcliffe Hospital, University of Oxford, Oxford, Great Britain

KEY WORDS: Pulmonary arteriovenous oxygen difference, pulmonary blood flow, spirometry.

The direct oxygen Fick approach is the standard reference technique for cardiac output measurement. It probably is the most accurate method currently available, although there are many possibilities of introducing errors, and considerable care is needed.

Essentials of the procedure[1–3]

The Fick method for the measurement of cardiac output using catheterization, is based on the theoretical principle enunciated by Adolph Fick in 1870. This principle states that the flow of blood in a given period of time is equal to the amount of substance entering the stream of flow in the same period of time divided by the difference between the concentrations of the substance in the blood upstream and downstream from its point of entry into the circulation. Pulmonary blood flow can be determined by measurement of oxygen uptake and the arteriovenous difference of oxygen across the lungs. In the absence of a shunt within the heart or between the great vessels, mean flow through the pulmonary circulation is the same as cardiac output, with a small error due to the presence of the bronchial circulation.

The estimation of cardiac output, which is usually expressed in $l\ min^{-1}$, thus requires measurement of oxygen uptake by the lungs and of the arteriovenous oxygen difference across the lungs. Cardiac output is calculated as:

oxygen uptake/arteriovenous oxygen difference.

OXYGEN UPTAKE

Oxygen uptake is assessed by measurement of oxygen extracted by the lungs from inspired air over a given period of time. This can be done using the closed-circuit method, in which the rate of oxygen usage is determined by the rate at which the rebreathing chamber of a standard metabolism apparatus, filled with oxygen, empties.

The open-circuit method is most frequently used. Here, oxygen uptake is measured as the difference between inspired and expired oxygen volume so that, in principle, both inspired and expired air volumes and oxygen concentrations have to be measured. A simpler procedure is to measure only the expired air volume, and to calculate the inspired air volume from the expired air volume and the amounts of oxygen and carbon dioxide it contains. Oxygen uptake can be assessed by the following equation:

$$\dot{V}O_2 = \dot{V}_E\,[0{\cdot}265 \times (1 - F_ECO_2) - 1{\cdot}265 \times F_EO_2]$$

in which: $\dot{V}O_2$ = oxygen uptake ($l\ min^{-1}$); $\dot{V}_E$ = expired air volume ($l\ min^{-1}$); F_ECO_2 = fractional amount of carbon dioxide in expired air (%); F_EO_2 = fractional amount of oxygen in expired air (%). In this equation the fractional amount of oxygen in the inspired air is taken as 0·2094.

By convention, the volume of gas exchanged is expressed as the volume the gas would occupy as a dry gas at standard temperature (0 °C or 273 K) and pressure (760 mmHg). These conditions are commonly abbreviated as STPD (Standard Temperature Pressure Dry) and obtained by multiplying the volume by:

$$\frac{P_A - P_A\,(H_2O)}{760\ \text{mmHg}} \times \frac{273}{273 + T_A}$$

in which: P_A = barometric pressure and $P_A\,(H_2O)$ = the water vapour pressure of saturated air at ambient temperature (T_A).

Correspondence: R. Fagard, M.D., Inwendige Ziekten–Cardiologie, Laboratorium voor Hartfunctie, U.Z. Pellenberg, Weligerveld 1, B-3041 Pellenberg, Belgium.

In the open-circuit method, the patient breathes room air through a three-port valve system and exhales through a tube so that the volume and the concentration of oxygen and carbon dioxide can be measured in the expired air. There are three data acquisition techniques.

The standard method for measurement of gas volumes is by spirometer and dry gas meter. Air can be collected in a Douglas bag and the volume pushed through a gas meter or into a spirometer, adjusting for temperature and humidity. Collection should begin and end at the same point in the respiratory cycle, preferably during inspiration. The fractional amounts of oxygen and carbon dioxide are then measured.

By the second method, expired air is passing through a mixing chamber so that mixed expired air can be continuously analysed. The mixing chamber has a volume of 5–10 l and contains several baffles to facilitate mixing. The larger the chamber, the more complete the mixing of expired gases, but the responsiveness to variations in gas composition is then greatly reduced because of the increased 'washout' time. The fractional amounts of oxygen and carbon dioxide are measured continuously on mixed air and integrated by a computer with the air volume measurement. The results can be displayed throughout the test and time-averaged data obtained afterwards.

A third approach is the breath-by-breath method. The flow of expired air is continuously measured. In addition, air is continuously removed from the valve at a constant rate and delivered to the oxygen and carbon dioxide analysers. Oxygen uptake is measured for each breath, and time-averaged values can be obtained by adding the values over a certain period of time.

There are several commercially available integrated systems that use a mixing chamber or the breath-by-breath method. Fractional amounts of oxygen and carbon dioxide can be measured by a variety of methods. Difficulties arise when air flow needs to be measured 'on-line' during the procedure. The most widely used detection device is the pneumotachograph, which is generally of reasonable accuracy for clinical use, although problems are caused by water vapour and expired air, which can condense upon the grids, changing the calibration in use. This is often overcome by increasing the temperature within the pneumotachograph, but this makes determining the exact temperature of the gas difficult. The size of the tachograph is also a problem; greater accuracy is achieved by using smaller devices at rest and larger ones during exercise and compromise is inevitable. There are other devices that depend upon changes in temperatures and a flow meter, but they have non-linear calibrations. Perhaps greater accuracy can be achieved using turbines. These are miniature turbine devices which appear to be accurate in measurement of flow over a wide range. Finally there are ultrasonic devices for the measurement of air flow, but experience of these is fairly limited.

ARTERIOVENOUS OXYGEN DIFFERENCE ACROSS THE LUNGS

The arteriovenous oxygen difference across the lungs is determined as the difference between the oxygen content of blood entering the lungs and that of blood leaving the lungs.

Mixed venous blood

The only segments of the circulation from which an adequate sample can be obtained of mixed venous blood prior to its entry into the lungs is the right ventricle and, preferably, the pulmonary artery; blood should not be sampled from the right atrium, where the various venous streams fail to mix completely. Sampling of blood from the pulmonary artery requires right-heart catheterization, which can readily be done without fluoroscopy with a floating catheter. The blood oxygen content of mixed venous blood changes with the phase of respiration[4]. Therefore, blood should be removed over a period of $\geqslant 5$ s to give an average rather than an instantaneous sample.

Systemic arterial blood

To estimate the oxygen content of blood leaving the lungs, blood is sampled from a systemic artery, and because the oxygen content of systemic arterial blood is commonly 2–5 $ml\,l^{-1}$ lower than that of blood leaving the alveoli, a small overestimation of cardiac output results.

Blood oxygen content

Usually 3–5 ml of blood are obtained in preheparinized syringes. The oxygen content in the blood can be measured by a variety of methods, several of them direct, such as the classic and accurate van Slyke vacuum extraction, and the manometric procedure, which is time consuming and requires considerable skill.

Most laboratories calculate oxygen content from measurements of oxygen saturation and haemoglobin concentration as the product of the haemoglobin concentration, the theoretical oxygen-carrying capacity of haemoglobin (1·34 ml O_2 g^{-1} haemoglobin) and the oxygen saturation of haemoglobin.

The arteriovenous oxygen difference can be calculated as:

haemoglobin concentration × 1·34 ×

(oxygen saturation of arterial − oxygen saturation of mixed venous blood).

In newer developments a fibreoptic system is added to the catheter to provide a continuous, 'real time' measurement of mixed venous oxygen saturation; arterial oxygen content can be obtained indirectly from ear oximeters, which are opto-electronic devices.

Limitations of the technique

The technique has limitations owing to its invasive nature, to practical and to theoretical errors associated with the sampling of gas and blood and with the various measurements, and is exact only if certain conditions are fulfilled[5,6].

(a) The technique requires right heart catheterization and sampling of arterial blood and is therefore not suitable for repeated measurements on different occasions.

(b) The method measures the mean flow over a period of time and is not suitable to follow rapid changes in flow from instant to instant.

(c) The intermittency of ventilation makes it necessary to collect the expired air for at least 1 min and ideally for about 3 min. During this time, it is believed that the rate of oxygen uptake measured at the mouth is the same as that occurring at the alveolar capillaries. The lungs comprise a large reservoir between the pulmonary airway and the alveolar membrane and, in certain circumstances, the amount of oxygen entering the blood each minute will differ from that entering the lungs. The volumes of oxygen and nitrogen in the lungs should therefore remain unchanged during the period of measurement, and this is not the case if there is a change in the rate of uptake of oxygen from the alveoli (e.g. at the beginning of exercise), if there is a change in the volume of ventilation, if the composition of inspired air is altered or if there is a change in the average total volume of gas held in the lungs.

(d) The Fick calculation will give a true measure of the average flow only if the flow is constant or if the oxygen contents of mixed venous and arterial blood are constant. The first condition is physiologically impossible because blood flow changes with the cardiac and respiratory cycles. In normal conditions, changes of oxygen content of end-capillary blood through the cardiac and respiratory cycles are unlikely to be substantial because this is at the plateau of the oxygen dissociation curve. There are small cyclical changes in oxygen content of mixed venous blood with the phase of respiration, which become more apparent on exercise.

(e) It follows that the Fick calculation of cardiac output assumes steady-state conditions. One should allow at least 2 min after a change in circulatory or respiratory conditions before making a determination of cardiac output, e.g. after a moderate increase of workload during exercise testing.

Comparison with other methods

The direct oxygen Fick technique is used as a standard reference method for measurement of cardiac output. It probably gives the most accurate measure of cardiac output, but a small error due to the presence of the bronchial circulation is inevitable. If absolute steady-state conditions were achieved then the Fick principle would be infallible, but there are many opportunities for error. In fact, the true value of cardiac output remains unknown. Therefore, comparison of other techniques with the Fick measurement is different from calibration and is essentially the comparison of a new method with an established reference technique rather than with the true quantity[7].

On the other hand, measurements of cardiac output by the direct oxygen Fick method have been compared with values obtained by measuring blood flow through the pulmonary artery directly with an optical rotatometer in anaesthetized dogs[8]. In 13 comparisons in 10 dogs, flow averaged $1{\cdot}01 \pm 0{\cdot}18$ (SD) l min^{-1} by the Fick method and $0{\cdot}98 \pm 0{\cdot}19$ l min^{-1} by the rotatometer; the difference between the Fick and the rotatometer method averaged $+ 23{\cdot}5 \pm 67{\cdot}8$ ml min^{-1} and the percentage deviation of the former from the latter $+ 2{\cdot}9 \pm 7{\cdot}0\%$.

Variability of results

AT REST

The reproducibility of cardiac output measurements by the direct oxygen Fick method in man has been studied by estimating cardiac output repeatedly within a short period of time (usually 15 min), leaving the right heart catheter in place.

Table 1 Median percentage difference of repeated measurements of oxygen uptake, arteriovenous oxygen difference and cardiac output in man at rest

Authors	Thomasson *et al.*[6]	Selzer *et al.*[7]	Fagard *et al.*[8,9]
Subjects	Healthy volunteers	Cardiopulmonary patients*	Hypertensive patients
Number	17	27	24
Age (years)	19–40	?	24–57
Gender (m/f)	11/6	?	17/7
Results:			
$\dot{V}O_2$	5·8%[a,c]	7·6%[a,d]	4·4%[b,e]
$d(AVO_2)$	5·2%[f]	7·2%[f]	5·2%[g]
CO	5·9%	8·5%	6·9%

Abbreviations: $\dot{V}O_2$: oxygen uptake; $d(AVO_2)$: arteriovenous oxygen difference: CO: cardiac output. [a]expired air volume measured by Tissot spirometer; [b]expired air volume measured by pneumotachograph; [c]fractional amount of oxygen measured by Haldane's apparatus; [d]fractional amount of oxygen measured by Scholander's method; [e]fractional amount of oxygen in inspired air and in expired air (from mixing chamber) measured by paramagnetic oxygen analysers; [f]oxygen content of blood measured by van Slyke's technique; [g]oxygen content of blood calculated from measurements of haemoglobin concentration and oxygen saturation. *Only the results of the patients with arterial O_2 saturation $\geq$90% and with normal cardiac output are given in the Table.

Table 1 summarizes the data for oxygen uptake, the arteriovenous oxygen content difference and cardiac output from four studies[9–12]. Reproducibility is given as the percentage difference between two repeated measurements, i.e. the ratio of the absolute difference divided by the average of the two values, expressed as a percentage. Only the median percentage difference was reported in one study[7], whereas this value could be calculated from individual data in the others; data from two reports[8,9] have been pooled.

The methodology used to measure oxygen uptake and arteriovenous oxygen difference varied between studies. The reproducibility of oxygen uptake was similar for data obtained through collection of air in a spirometer and for the mixing chamber technique. Similarly, the reproducibility of the arteriovenous oxygen difference was comparable when calculated from haemoglobin concentration and oxygen saturation or when measured directly. The median percentage difference of cardiac output was 5·9%, 8·5% and 6·9% in three studies. It should be realized that direct comparison between the various techniques cannot be made because they have been collected on different subjects in different laboratories.

ON EXERCISE

The reproducibility of cardiac output measured by the direct oxygen Fick method can be assessed using data reported by Fagard *et al.*[13]. Ten hypertensive patients (eight men), aged 43 ± 12 (SD) years performed two graded uninterrupted submaximal bicycle exercise tests with a rest period of 75 min between the tests. Exercise was started at 20 W, and workload was increased by 30 W every 4 min, up to the ventilatory anaerobic threshold. The highest workload averaged 101 ± 19 W. Oxygen uptake was measured using the mixing chamber technique and the blood oxygen content was calculated from haemoglobin concentration and oxygen saturation. There were no systematic differences between the two tests: at 50 W, cardiac output averaged $9{\cdot}4 \pm 1{\cdot}1\ l\,min^{-1}$ at the first test and $9{\cdot}1 \pm 1{\cdot}5\ l\,min^{-1}$ at the second; these values averaged 12·9 ± 1·9 and $12{\cdot}7 \pm 2{\cdot}0\ l\,min^{-1}$, respectively, at the highest workload. At 50 W, the median percentage difference averaged 4·2% for oxygen uptake, 5·8% for the arteriovenous oxygen difference, and 5·4% for cardiac output; these values averaged 3·5%, 2·6% and 3·0%, respectively, at the highest workload.

Practical recommendations

(1) Steady-state must be achieved before measuring cardiac output. A period of at least 15 min should be allowed after catheterization. At least 2 min should elapse after a change in the level of physical activity.

(2) Oxygen uptake should be measured over 3–4 min at rest and 1–2 min during exercise.

(3) Midway in the measurement period of oxygen uptake, 3–5 ml of arterial blood and of mixed venous blood (from the pulmonary artery) should be drawn slowly (>5 s) into preheparinized syringes for measurement of oxygen content. This can be done by measuring haemoglobin concentration and oxygen saturation, or by the more tedious van Slyke method.

The authors gratefully acknowledge the secretarial assistance of N. Ausseloos.

References

[1] Guyton AC, Jones CE, Coleman TG. Measurement of cardiac output by the direct Fick method. In: Circulatory physiology: Cardiac output and its regulation, IInd edition. Philadelphia: W.B. Saunders Company 1973; 21–39.
[2] Harris P, Heath D. The measurement of flow. In: The human pulmonary circulation. Edinburgh: Churchill Livingstone 1977; 78–96.
[3] Janicki JS, Shroff SG, Weber KT. Instrumentation for monitoring respiratory gas exchange. In: Cardiopulmonary exercise testing: physiologic principles and clinical applications. Weber KT, Janicki JS, Eds. Philadelphia: W.B. Saunders Company 1986; 113–25.
[4] Wood EH, Bowers D, Shepherd JT, Fox IJ. Oxygen content of 'mixed' venous blood in man during various phases of the respiratory and cardiac cycles in relation to possible errors in measurement of cardiac output by conventional application of the Fick method. J Appl Physiol 1955; 7: 621–8.
[5] Visscher MB, Johnson JA. The Fick principle: analysis of potential errors in its conventional application. J Appl Physiol 1953; 5: 635–8.
[6] Stow RW. Systematic errors in flow determinations by the Fick method. Minnesota Med 1954; 37: 30–5.
[7] Bland JM, Altman DG. Statistical methods for assessing agreement between two methods of clinical measurement. Lancet 1986; i: 307–10.
[8] Seely RD, Nerlich WE, Gregg DE. A comparison of cardiac output determined by the Fick procedure and a direct method using the rotatometer. Circulation 1950; 1: 1261–6.
[9] Thomasson B. Cardiac output in normal subjects under standard basal conditions. The repeatability of measurements by the Fick method. Scand J Clin Lab Invest 1957; 9: 365–76.
[10] Selzer A, Sudrann RB. Reliability of the determination of cardiac output in man by means of the Fick principle. Circ Res 1958; 6: 485–90.
[11] Fagard R, Amery A, Reybrouck T, Lijnen P, Billiet L. Acute and chronic systemic and pulmonary hemodynamic effects of angiotensin converting enzyme inhibition with captopril in hypertensive patients. Am J Cardiol 1980; 46: 295–300.
[12] Fagard R, Staessen J, Amery A. The use of Doppler echocardiography to assess the acute haemodynamic response to felodipine and metoprolol in hypertensive patients. J Hypertens 1987; 5: 143–9.
[13] Fagard R, Bulpitt C, Lijnen P, Amery A. Response of the systemic and pulmonary circulation to converting enzyme inhibition (captopril) at rest and during exercise in hypertensive patients. Circulation 1982; 65: 33–9.

The dye dilution method for measurement of cardiac output

P. LUND-JOHANSEN

Section of Cardiology, Medical Department, University of Bergen, School of Medicine, Bergen, Norway

KEY WORDS: Haemodynamics, dye dilution, cardiogreen, densitometers.

The dye dilution method for measuring cardiac output is based on injecting rapidly a known quantity of a dye at one site into the circulatory system, and withdrawing blood at a distal site for determination of a concentration curve of the dye.

Flow (Q) is calculated by the formula:

$$Q = \frac{m}{\bar{c} \times t}$$

where m is the amount of dye injected, $\bar{c}$ mean concentration of dye and t the time of the concentration curve without recirculation.

In recent years the only dye used has been indocyanine green (cardiogreen) which has its absorption maximum in the infrared part of the spectrum (at 805 μm) – where oxyhaemoglobin and reduced haemoglobin transmit light equally.

Several densitometers for cardiogreen have been developed. The Christian Michelsen Institute densitometer used in our laboratory was found to give very accurate measurements (error $< \pm 2\%$) of blood flow in model experiments, for flows ranging from 2 to 12 l min^{-1}. The more modern densitometers are usually equipped with computers.

The cardiogreen method is probably one of the most accurate methods to study cardiac output during exercise. The error of a single determination of cardiac output values at rest and during exercise is less than $\pm 5\%$. The method does not allow measurement of 'beat to beat' changes, and requires a cardiac output which is stable for approximately 10 s during exercise and 30 s at rest. It has been extensively used in our laboratory to study changes in central haemodynamics in essential hypertension at rest and during exercise, and also to study the haemodynamic alterations induced by anti-hypertensive agents.

The method is safe and may be used on an outpatient basis. Another advantage is that the arterial catheter can be used for determination of blood pressure during exercise, when all methods based on arm cuff sometimes give very inaccurate values, particularly for diastolic blood pressure.

Introduction

The ideal method for measurement of cardiac output should be simple enough to be used in outpatients and also accurate over a large flow range – from low cardiac outputs of 2–3 l min^{-1} in heart failure, to high cardiac outputs of 20 to 30 l min^{-1} during intense exercise in healthy subjects. Ideally, the method should also be rapid and suitable for computer analysis. The dye dilution method fulfils these criteria, and represents probably one of the most accurate methods for determining cardiac output at high levels of physical exercise. An obvious drawback is that it is invasive.

The principle of the method is to inject rapidly a known quantity of a dye at one site into the circulatory system and withdraw blood at a distal site for determination of concentration of dye. Flow (Q) is calculated

$$Q = \frac{m}{\bar{c} \times t}$$

Address for correspondence and reprint requests: Prof. P. Lund-Johansen, Medical Department, N-5021 Haukeland Hospital, Bergen, Norway.

(m = injected amount of dye, $\bar{c}$ = mean dye concentration, t = time of concentration curve).

Historical background

A detailed review of this area was published by Fox in 1962[1]. As early as 1827, Hering determined circulation time in a horse by injecting potassium ferrocyanide into the jugular vein and sampling from the opposite side. However, it was Stewart who first measured cardiac output by injecting an indicator (hypertonic saline) and recording the concentration at a distal site.

Many years later, Hamilton and his group[2] modified Stewart's single injection method and discovered how the recirculation problem could be overcome by semilogarithmic replotting and linear extrapolation of the down-slope of the original curve. Subsequently, the method was referred to as the 'Stewart-Hamilton technique'. Hamilton originally used T 1824 (Evans blue) and, in 1948, he published a paper on the comparison of the Fick and dye injection methods of measuring cardiac output in man. He found an agreement or more than 30% which he considered acceptable. The following year, Werkø *et al.*[3] published a modification of the method which gave a better correlation between dye and Fick methods.

However, Fox and Wood[4] demonstrated that T 1824 could not be detected accurately if the oxygen concentration in blood fluctuated. This resulted in the development of the dye indocyanine green (later called cardiogreen)[4].

In contrast to Evans blue, methylene blue and indigocarmine, which all preferentially absorb light in the red region of the spectrum, indocyanine green has its peak absorption at 800 μm, where oxyhaemoglobin and reduced haemoglobin transmit light equally. Cardiogreen has become established as the only dye used, and instruments have been developed to measure the concentration of cardiogreen in blood drawn through a cuvette or in the blood flowing through the pinna of the ear (see later).

Theoretical basis of the dye dilution method of measuring cardiac output

The theoretical validity of the dye dilution method in humans has been discussed at length, and an excellent review has been given by Zierler[5]. In practice, a known quantity of a dye is injected into a flow of unknown rate, and the concentration of dye is then measured distal to the injection site. Prerequisites for the validity of the method are that the flow and volume of the vascular bed are constant, and every unit of fluid entering the system must leave it.

The calculation of flow by the formula

$$Q = \frac{m}{\bar{c} \times t}$$

implies that the area under the curve without recirculation can be defined. In low flow situations, and particularly if the dye is injected peripherally, it becomes impossible to reconstruct the straight line of the exponential fall in the concentration of the dye, and the method is invalid.

Zierler[5] has pointed out that the equation is valid also in systems that do not have a single inflow and a single outflow orifice. Violation of the theory by pulsatile flow is not critical if the alterations in transit time fluctuate rapidly.

Theoretically, injection must be 'instantaneous', but this is clearly impossible. If the mean transit time through the system is long compared with the mean time of injection, it is usually acceptable to ignore this violation of the theory.

Indocyanine green (cardiogreen)

Cardiogreen (Hynson, Westcott & Dunning, Baltimore) is the only dye used to any extent for determination of cardiac output over the last three decades. It is remarkable that a compound has remained the first choice for a diagnostic purpose for so many years. There are several reasons for this. As already stated, indocyanine green has its absorption maximum in the infrared part of the spectrum (805 μm) – where the absorption spectrum of oxyhaemoglobin and unsaturated haemoglobin cross. Cardiogreen is rapidly excreted and at least 10 cardiac outputs can be recorded within 1 h, provided that the densitometer is linear in concentrations up to 20 mg^{-1} l and conventional doses are injected.

Cardiogreen seems to be completely non-toxic, as originally reported by Fox and Wood[4]. In our laboratory we have been using cardiogreen for 25 years and have never observed any reactions to the compound.

Nevertheless, indocyanine green must be handled correctly. The powdered substances must be diluted in the accompaning sterile solution. The solution should be protected from light and air. At the concentrations used for measurements of cardiac output the solution is stable at room temperature for more than 8 h, and the same stock of solution should be used for calibration and injection into the patient. The dye undergoes changes in spectral absorption when added to blood. This process is however quite fast, and if the time from injection to withdrawal is more than 5 s, there is good agreement between the dye curves derived using different injection sites[6].

Cuvette densitometers for measurement of cardiogreen

During the last thirty years, several densitometers for cardiogreen have been developed. As the principles are basically the same, the cuvette densitometer we have been using in our laboratory will be given as an example[7].

The instrument consists of a cuvette which can be pushed in and out of a light-proof cuvette box. Blood flow through the cuvette is maintained constant by means of a motordriven pump which can be set at various speeds. The usual withdrawal speed is 15 ml min^{-1}, and slower speeds should not be used, particularly during exercise. The photoelectric equipment consists of four photo-cells (cadmium sulphide photoconducted cells covered by a gelatin filter) (Wratten 87 og 88 A) with a maximum sensitivity at about 8000 A. The photo-cells are coupled in a Wheatstone bridge. All four cells receive light from the same source. A detailed description of the instrument is found in reference 7.

When the instrument is used, blood is drawn through the cuvette and the instrument is calibrated so the voltage output to the recorder is zero from both photo-cell systems. During testing of the instrument, it was shown that blood of various degrees of saturation did not effect stability. The apparatus was also insensitive to changes in temperature from 20 to 38 °C.

For calibration of a curve, exactly 10 ml blood are taken from the patient, and mixed with 50 μg dye using a micropipette to give a concentration of 5 mg l^{-1}. The apparatus is set at zero with 15 ml min^{-1} of the patient's blood drawn through the cuvette, and then the calibration blood is incorporated. Thereafter blood from the patient is drawn through again, the dye is injected and the dilution curve recorded. Before the second dye injection, the apparatus is readjusted to zero.

The advantages of the Wheatstone bridge circuit with cadmium-sulphide resistant cells, as used in our instrument, are a favourable voltage difference for minor concentrations of cardiogreen, and good compensation for variations in the light source voltage and temperature. The instrument is linear up to 10 mg l^{-1} of cardiogreen.

Model experiment to control the dye dilution instruments for determination of cardiac output

For this purpose, a circulatory model was constructed[8]. The 'heart' was a plastic pump driven by an electromagnetic motor. The frequency and stroke volume could be regulated within the ranges appropriate for the human heart, up to a flow of 12 l min^{-1}. The pump was connected to rubber tubes of 16 mm internal diameter and a reservoir with a total volume of approximately 5–6 l (normal human blood volume). The system was filled with heparinized filtered ox blood. In order to obtain accurate measurements of the flow, a valve was constructed and connected to an electrical stop watch. When the flow was switched from the circulating system to a measurement cylinder (to obtain the exact volume) the precise time was measured and the flow could be calculated from the volume in the cylinder and the time.

Simultaneous control measurements of flow during dye injection would eliminate the recirculation and, in addition, require larger quantities of blood than usually obtained from one ox. In order to control the stability of the flow during dye injections a flowmeter (Meter-Flow Ltd, Feltham, England) with a counter was introduced into the system. This showed flow variation with an accuracy of within ± 10 ml min^{-1} throughout the flow range[8].

A polyethylene catheter of the same type as that used in humans was introduced in the afferent tube ('large vein') and connected to the injection device used in human experiments. Another catheter was introduced into an efferent tube ('artery'). This catheter was connected to a three-way tap, one lumen leading to a strain gauge transducer, the other to the densitometer. The speed of the withdrawal pump was 15 ml min^{-1} (the same as in human experiments). The circulatory model was set at different flows between 2 and 12 l min^{-1}. At each flow level, three or four curves were registered. A typical dye curve from the model experiment is shown in Fig. 1. The results showed that at both high and low flow rates the differences between the measurements of flow by the dye dilution method and the true flow were very small. The greatest difference was less than $\pm 3\%$ of true flow, as seen in Table 1. In conclusion, the Christian Michelsen's Institute densitometer was found to give a reading

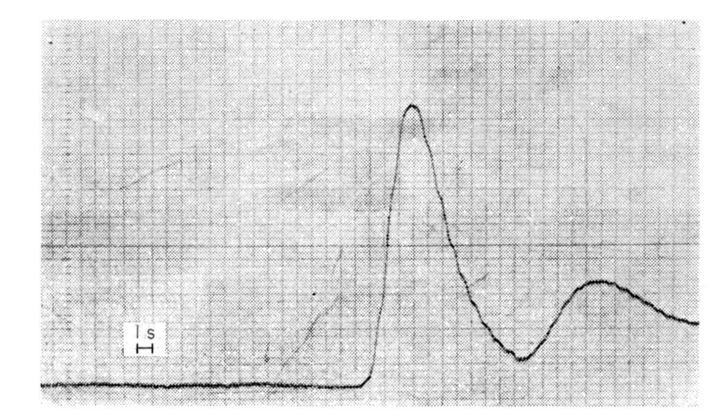

Figure 1 Typical dye-dilution curve from model experiment (from [8]).

Table 1 Accuracy of dye dilution method in model experiments. Single determinations. High flow rates (3·8–11·7 l min⁻¹). (From [8])

Flow by flowmeter	3·85	3·87	3·86	6·36	6·38	6·25
Flow by dilution method	3·82	3·90	3·82	6·34	6·34	6·42
Difference	−0·03	+0·03	−0·04	−0·02	−0·04	+0·17
Greatest difference (%)						+2·7

Flow by flowmeter	8·51	8·49	8·52	8·60	11·5	11·4	11·7	11·7
Flow by dilution method	8·34	8·41	8·50	8·62	11·2	11·1	11·9	11·9
Difference	−0·17	−0·08	−0·02	+0·02	−0·3	−0·3	+0·2	+0·2
Greatest difference (%)						−2·6		

of blood flow very close to actual blood flow, determined by measurement cylinder and stop watch. The instrument has been used for more than 20 years.

Other densitometers

During the last few decades several other densitometers have been constructed (i.e. Nilson[9] and Waters[10]). As the Waters is probably most widespread, a brief description will be given[10]. The Waters DC-410 densitometer transducer consists of a dual cell optical system with two light-emitting diodes. One cell has an optical system with a peak response at the wavelength corresponding to maximum light absorption by cardiogreen (800 μm). The other photo-cell has an optical system which permits adjustment for changes in flow, carbon dioxide pressure, a haematocrit and other non-specific factors.

A recorder measures the electrical signal received from the control unit. As the dye flows through the cuvette between the light sources and the photocells, the change in light transmission to the photo-cells changes the resistance of the photo-cells in direct relation to dye density in the optical part. This change in resistance produces a change in voltage. The voltage is amplified so that the signal output is suitable for the recorder. The system is provided with a cardiac output computer and also equipped for the dynamic calibration method.

Calibration methods

THE 'DYNAMIC' CALIBRATION METHOD

A quicker calibration method than conventional dilution[8] (see previous section) is the 'dynamic' method[11]. An additional component is inserted between the arterial line and the cuvette, and this allows a small quantity of dye to be injected and produce a curve of similar amplitude to the curve obtained from the patient. The 'true flow' is the same as that generated by the withdrawal pump. A known quantity of cardiogreen (usually 10–20 μl of the same dye solution as used for the patient's curve) is injected through the rubber tube using a 50 μl Hamilton syringe. The rate of the flow through the system is similar to that used for obtaining patient data, and is measured accurately by timing the flow of 20 or 30 ml min^{-1} into the withdrawal syringe with a stop watch.

As the flow in the calibration system has been measured, the flow in the patient (cardiac output) can be obtained by comparison:

$$Q = Q_C \times \frac{A_C}{I_C} \times \frac{I}{A}$$

Q_C = flow in calibration system. I_C = dye injected in calibration system. A_C = curve area in calibration system. I = dye injected into patient. A = area of patient curve.

One advantage of this method is that it can be used with cardiac output computers (see later).

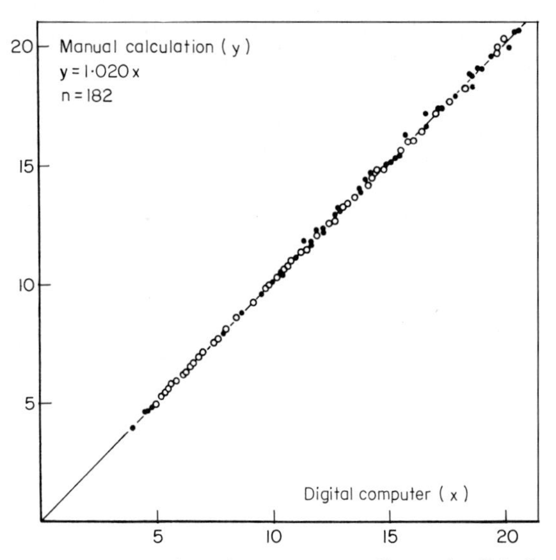

Figure 2 Relation between manually and digitally computed values of cardiac outputs (from [14]).

CALCULATION OF THE DYE DILUTION CURVES

(*The extrapolation – the area without recirculation*)

The original Stewart-Hamilton method involved replotting the curve onto semilogarithmic paper, then drawing a straight line and thus calculating the area under the curve. This is a very time-consuming procedure and several simplification methods have been tried over the years.

Dow[12] introduced the so-called 'Forward-Triangle method'. An empirical area formula was developed which depended only upon appearance time, peak concentration time and peak concentration, Schad *et al.*[13] suggested a very simple calculation method. They measured peak concentration (H) and the distance (t_{75}) between the time of indicator appearance and the time of the decay of indicator concentration to 75% of peak concentration: $A = H \times t_{75}$. The mean difference between cardiac output calculated from such empirical areas and from extrapolation planemetry was 6%. The authors recommended that the method should not be employed if high precision were required.

As an alternative to the conventional manual method and these crude calculations[12,13], we developed a digital computer method[14]. A programme was written which calculated the area from t = 0 to t = tn using Simpson's rule, and the remaining area according to a calculated formula[14]. We investigated 182 dye curves from 20 male hypertensive patients, and the relationship between manually calculated and digitally computed values of cardiac outputs are shown in Fig. 2. As can be seen, the correlation was extremely good over a large flow range, the correlation coefficient was 0·9996 and the estimated standard error was 0·13 l min^{-1}.

An attempt to use an analogue computer gave greater disagreement (standard errors 0·67 l min^{-1})[14], but was quicker and could be used in routine clinical work.

Other instruments developed during the last decade are normally supplied with a computer (e.g. Waters instrument[10]). The author has not had access to the construction of these, and is not aware of systematic studies of their accuracy. Ideally, the whole set-up should be tested in a circulatory model, or compared with a system where this has been done.

Methodological errors

To obtain reliable determination of cardiac output in man (and animals) several precautions must be taken. Calibration should be performed before taking recordings in new situations, such as after a change of posture, after drug intervention, during exercise testing etc. When several injections are performed over a short time, there is a build-up of concentration of dye. In such circumstances a calibration procedure with 5 and 10 mg l^{-1} should be used. It is important that the injection is rapid and central. Peripheral injection may cause curves that are too flat and impossible to use. This is particularly true when cardiac output is low[15,16], but less critical during exercise.

Obviously great care must be taken to avoid any leakage of dye. The withdrawal system must be stable and the system air-proof, in order to avoid micro-bubbles. The method is invalid if there are shunts (i.e. atrial septum defect) or valvular lesions.

When all technical errors were avoided, the error of single determination of CO values at rest and during exercise was found to be less than $\pm 5\%$, calculated from the formula

$$\sqrt{\frac{(d^2)}{2n}}$$

where d is the difference between duplicate determinations and n the number of duplicate determinations.

Ear-piece densitometer

The most accurate densitometers for measurement of cardiac output are those based on withdrawal of blood through a cuvette densitometer. The obvious drawback of this method is that it is invasive, and hence use of the pinna of the ear has been tried instead[17,18].

In principle this method assumes that, when blood flow through the ear pinna is stable and the ear-piece densitometer is set to zero, it should read accurately and linearly the concentrations of cardiogreen through the ear. The densitometer is connected to a transducer with a recorder, and dye dilution curves are constructed similar to those obtained by using the conventional blood cuvette.

One problem with the ear-piece densitometer is, of course, calibration. Calibration of the ear curves assume linearity of deflection for dye concentration. The deflections from baseline at 90–100 s after each dye injection are assumed to be due to the concentration of indocyanine green, determined independently by a Beckman spectrophotometer in blood samples withdrawn simultaneously from the radial artery or from the brachial vein.

In order to obtain accurate results, a dichromatic ear-piece instrument must be used. The standard deviation of differences has been reported to be 14%[17,18]. There is little information on the accuracy of the ear-piece densitometer during physical exercise.

Discussion and conclusions

The cardiogreen method has been used to determine cardiac output for more than 25 years. Provided there is a linearly responding densitometer and injection of the dye cardiogreen in the region of the subaxillary vein or more centrally, this technique seems to give a surprisingly accurate cardiac output measure provided no shunts or valvular lesions are present. Our own model experiments showed that the disagreement between true flow and flow assessed by dye was only $\pm 2\%$[8]. The method requires a relatively stable cardiac output, since 10–30 s elapse from injection to completion of the dye curve. In man, the error of single determination was less than $\pm 5\%$. The method will of course not show 'beat to beat' variations.

A drawback with the method is that it requires an intra-arterial catheter and, in order to obtain a sufficiently high flow rate, the brachial artery is usually used. The method can be performed on an outpatient basis as it is not necessary to enter the heart.

The dye dilution cardiac output method has been less popular during the last 10 years, following the development of the Swan-Ganz catheter and the thermodilution method[19–21]. By the thermodilution method, cardiac output can be measured very quickly and it is possible to determine pulmonary artery pressures and pulmonary wedge pressure. For these reasons, the thermodilution method using the Swan-Ganz catheter has become very popular and has replaced the dye dilution method in most laboratories. Several studies have shown good agreement between cardiac output by dye and thermodilution at rest[22,23].

In our laboratory we have been particularly interested in haemodynamics in hypertension at rest and during exercise, when accurate measurement of blood pressure is also of great importance. During exercise it is impossible to obtain reliable values of diastolic blood pressure without placing a catheter in a large artery (i.e. the brachial artery). We have tried many of the new non-invasive automatic instruments for 24 h measurement of blood pressure and compared the results with intra-arterial recordings at rest as well as during exercise. The results show that even when the arm with the cuff is relaxing on a pillow during bicycle exercise, we sometimes see large deviations between the intra-arterial pressure and cuff pressure[24]. Clearly, for studies of central haemodynamics in hypertensive subjects, the dye dilution method seems to be a logical choice — especially since it also allows intra-arterial blood pressure to be recorded. The method has been used in our laboratory for 25 years. It is safe, has permitted serial observations over as many years[26] and also has proved useful for studies of the many haemodynamic effects of antihypertensive agents[27].

An accurate computer programme will, of course, save a tremendous amount of time in the otherwise laborious procedures of manually calculating dye dilution curves. However, when demonstration of minor changes in the circulatory system are important, it is not possible to compromise on the accuracy of the methods to be used. With this in mind, the cardiogreen dilution method should continue to be used in the future, particularly for the study of haemodynamics during exercise.

References

[1] Fox IJ. History and developmental aspects of the indicator-dilution technic. Circ Res 1962; 10: 181–92.

[2] Hamilton WF, Riley RL, Attyah AM *et al.* Comparison of the Fick and dye injection methods of measuring the cardiac output in man. Am J Physiol 1948; 153: 309–32.

[3] Werkö L, Langerlöf H, Bucht H, Wehle B, Holmgren A. Comparison of the Fick and Hamilton methods for the determination of cardiac output in man. Scand J Clin Lab Invest 1949; 1: 109–13.

[4] Fox IJ, Wood EH. Indocyanine green: Physical and physiologic properties. Staff meetings of the Mayo Clinic 1960; 35: 732–44.

[5] Zierler KL. Theoretical basis of indicator-dilution methods for measuring flow and volume. Circ Res 1962; 10: 393–407.

[6] Saunders KB, Hoffman JIE, Noble MIM, Domenech RJ. A source of error in measuring flow with indocyanine green. J Appl Physiol 1970; 28: 190–8.

[7] Knutsen B, Lund-Johansen P, Hatletveit E. A linearly responding monochromatic cuvette densitometer for indocyanine green. Scand J Clin Lab Invest 1966; 18: 673–8.

[8] Lund-Johansen P. Hemodynamics in early essential hypertension. Acta Med Scand 1967; 181 (Suppl); 482: 1–100.

[9] Nilson NJ. A linearity responding dichromatic cuvette densitometer for dye-dilution curves. Scand J Clin Lab Invest 1963; 15 (Suppl 69): 181–92.

[10] Waters Instruments Inc. Co-10 Cardiac output computer Instruction Manual. Rochester; 1988.

[11] Shinebourne E, Fleming J, Hamer J. Calibration of indicator dilution curves in man by the dynamic method. Br Heart J 1967; 29: 920–5.

[12] Dow P. Dimensional relationships in dye-dilution curves from humans and dogs, with an empirical formula for certain troublesome curves. J Appl Physiol 1955; 7: 399–408.

[13] Schad H, Brechtelsbauer H. Eine zeitsparende, empirische Methode zur Berechnung der Flächen von Farbstoffverdünnungskurven bei der Bestimmung des Herzminutenvolumens. Anaesthesist 1981; 30: 243–5.

[14] Mørkrid L, Lund-Johansen P. Comparison of analog and numerical computation of cardiac output from dye dilution curves. Acta Med Scand 1977; 603 (Suppl): 15–21.

[15] Carey JS, Hughes RK. Cardiac output. Clinical monitoring and mangement. Am J Thor Surg 1969; 7: 150–76.

[16] Chamberlaine JH. Cardiac output measurement by indicator dilution. Biomed Engin 1975; 10: 92–7.

[17] Reed JH, Wood EH. Use of dichromatic earpiece densitometry for determination of cardiac output. J Appl Physiol 1967; 23: 373–80.

[18] Robinson PS, Crowther A, Jenkins BS, Webb-Peploe MM, Coltart DJ. A computerised dichromatic earpiece densitometer for the measurement of cardiac output. Cardiovasc Res 1979; 13: 420–6.

[19] Ehlers KC, Mylrea KC, Waterson CK, Calkins JM. Cardiac output measurements. A review of current techniques and research. Annals Biomed Engin 1986; 14: 219–39.

[20] Kadota LT. Theory and application of thermodilution cardiac output measurement: A review. Heart Lung 1985; 14: 605–16.

[21] Rubin Rubin SA, Siemienczuk D, Nathan MD, Prause J, Swan HJC. Accurary of cardiac output, oxygen uptake, and arteriovenous oxygen difference at rest, during exercise, and after vasodilator therapy in patients with severe, chronic heart failure. Am J Cardiol 1982; 50: 973–8.

[22] Weisel RD, Berger RL, Hechtman HB. Measurement of cardiac output by thermodilution. New Engl J Med, 1975; 292: 682–4.

[23] Runciman WB, Ilsley AH, Roberts JG. An evaluation of thermodilution cardiac output measurement using the Swan-Ganz catheter. Anaesth Intens Care 1981; 9: 208–20.

[24] White WB, Lund-Johansen P, Omvik P. Assessment of four ambulatory blood pressure monitors and measurements by clinicians versus intraarterial blood pressure at rest and during exercise. Am J Cardiol 1990; 65: 60–6.

[25] Lund-Johansen P, Omvik P. Hemodynamic patterns of untreated hypertensive disease. In: Laragh JH, Brenner BM, eds. Hypertension: pathophysiology, diagnosis, and management. New York: Raven Press, 1990: 305–27.

[26] Lund-Johansen P. Central haemodynamics in essential hypertension at rest and during exercise: A 20-year follow-up study. J Hypertens 1989; 7 (Suppl 6): S52–S55.

[27] Lund-Johansen P. Hemodynamic effects of antihypertensive agents. In: Doyle AE, ed. Handbook of hypertension, Vol 11. Clinical pharmacology of antihypertensive drugs. Amsterdam: Elsevier, 1988: 41–72.

Some comments on the usefulness of measuring cardiac output by dye dilution

P. W. de Leeuw and W. H. Birkenhäger

Department of Medicine, Zuiderziekenhuis, Rotterdam, The Netherlands

KEY WORDS: Dye dilution, hypertension.

In this paper, a few aspects of the dye dilution method for measuring cardiac output are described. Data obtained with this technique show good reproducibility, although any first determination tends to be higher than subsequent ones. The method has been found to be useful in the assessment of cardiac function in, for instance, hypertensive patients. An additional advantage is that it can be used also in exercising subjects. As the dye dilution technique still is considered to be the standard method, other methods, such as thermodilution, should always be validated in comparison with the dye dilution technique.

Introduction

Proper evaluation of the cardiovascular system, including determination of cardiac output, is often essential in circulatory disease states. Although routine measurement of cardiac performance in patients with hypertension is not necessary, data on this variable may help in understanding the basic haemodynamic fault present in that disorder.

In recent years, several non-invasive techniques to assess cardiac output have been developed, some of which have gained tremendous popularity. However, it should not be forgotten that the standard method still is the dye dilution method. Elsewhere in this issue Lund-Johansen describes the background and the validation of this technique[1]; here we add data on reproducibility as well as on comparisons with some other methods, notably the thermodilution technique. Finally, a few comments will be made with respect to clinical applications.

Reproducibility of the method

Attempts to estimate the reproducibility of the dye dilution method within a single session are invariably associated with problems of interpretation. Indeed, any scatter in the data reflects not only true variations in the results of the measurements but also the impact of the physiological instability of cardiac output over even a short period of time. Despite this theoretical drawback we have been able to obtain reasonably comparable data with cardiac output measured at intervals of 10 min. The figures presented in Table 1 clearly show that initial determinations tend to yield higher values for cardiac output than subsequent ones. When we systematically discarded the results from the first measurements we found the variation of repeated determinations to be within 8%, which seems to be acceptable. However, the same data indicate that, with time, cardiac output may vary considerably and also unpredictably. In other words, both increases and decreases may be found without apparent reasons.

Data on the agreement between two consecutive measurements of cardiac output have been reported mainly in the older literature. For instance, in 60 subjects, Brandfonbrener and associates found an estimated error of 6·4% for a single observation; in the same individual, the difference between two determinations averaged 7·9% of the mean of the pair[2]. Interestingly, these authors also found the

Table 1 Results of cardiac output determinations (dye dilution), taken serially at 10-min intervals (n = 15)

Time (min)	Cardiac output ($l\,min^{-1}$)
0	7·85 ± 1·90
10	7·18 ± 1·38
20	6·99 ± 1·52
30	7·36 ± 1·80
40	7·30 ± 1·11

Address for correspondence: P. W. de Leeuw, M.D., Dept. of Medicine, Zuiderziekenhuis, Groene Hilledijk 315, 3075 EA Rotterdam, The Netherlands.

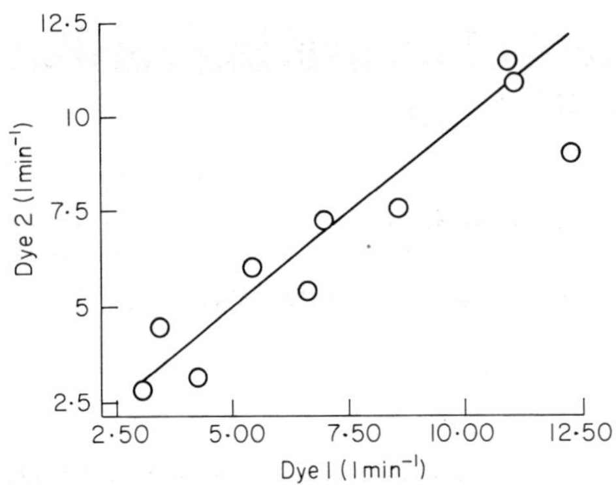

Figure 1 Reproducibility of two consecutive determinations of cardiac output by dye dilution: simple scatterplot. Line of identity has been drawn.

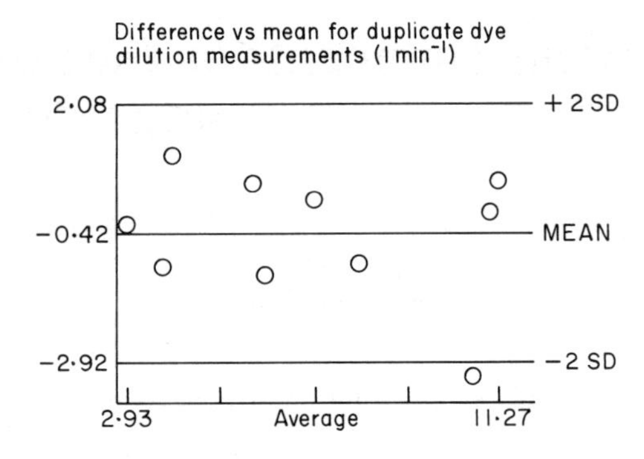

Figure 2 Reproducibility of two consecutive determinations of cardiac output by dye dilution: plot of the difference between the two determinations against their average value. The mean difference (second minus first) is $-0{\cdot}42\,l\,min^{-1}$ (middle horizontal line). The upper and lower horizontal line mark two standard deviations of this mean.

mean value of the first determination to be slightly higher than that of the second. Using bromsulphalein as the indicator, Sannerstedt found an error of 9·6% for a single determination of cardiac output[3]; again, second readings were lower than initial ones. Recently, Russell and coworkers found a very good agreement between duplicate measurements, the coefficient of repeatability being only $0{\cdot}88\,l\,min^{-1}$[4].

Given the invasive nature of the investigation, it is usually not possible to obtain repeated measurements on separate occasions in the same subjects. However, we have been able to do so in 10 patients with essential hypertension who participated in drug trials. Having acted as untreated controls for 2 weeks, they all volunteered to undergo a second examination under exactly similar conditions as the first. During the first evaluation, their cardiac output averaged $7{\cdot}26 \pm 3{\cdot}34\,l\,min^{-1}$ (mean ± standard deviation); it was $6{\cdot}85 \pm 3{\cdot}05\,l\,min^{-1}$ on the second day. A simple scatterplot of the data showed reasonable agreement between the two measurements (Fig. 1). When we plotted our results more appropriately, as suggested by Bland and Altman[5], the picture shown in Fig. 2 emerges. It now appears that there is no relation between the differences and the height of cardiac output and that the two determinations differ, on average, by $420\,ml\,min^{-1}$. Moreover, most of the differences will not exceed $1{\cdot}25\,l\,min^{-1}$. Finally, these data again demonstrate a tendency for the second assessment to yield lower values.

Comparison with other techniques

The dye dilution technique has been compared with most other methods for measuring cardiac output. Although reasonable correlations have been described with non-invasive methods such as echocardiography, impedance cardiography and echo-Doppler, one should realize that such comparisons are not entirely valid. These methods measure stoke volume and not cardiac output; the dye dilution technique, on the other hand, records mean flow over many seconds and 'averages' this to minute volume. This fundamental problem alone could account for a large part of the differences in results. Our data on impedance cardiography, which are published elsewhere in this issue, may exemplify this[6]. Likewise, Alpert and coworkers did not find any significant correlation when outputs obtained by dye dilution or echocardiography were compared; only when values were 'normalized' for heart rate (i.e. stroke volume) a significant, though still very poor, correlation ($r = 0{\cdot}76$) emerged[7].

For the present discussion, however, it seems more relevant to highlight the comparison of the dye dilution method with another averaging technique; here we will restrict ourselves to thermodilution. The latter is relatively easy to use and amenable to automation. Previous studies have shown close correlations between both methods in terms of correlation coefficients[8,9], but as we realize now, such presentations conceal important information[5]. In a series of 16 patients with uncomplicated essential hypertension, we have compared thermodilution with dye dilution by measuring cardiac output with both methods at exactly the same moment. Our data emphasize the serious overestimation of cardiac output by thermodilution.

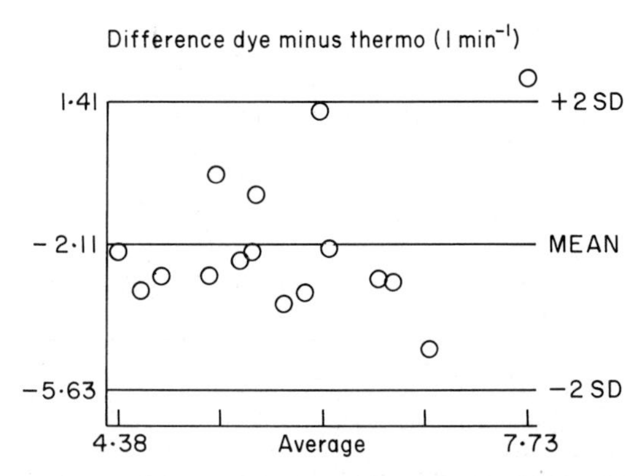

Figure 3 Comparison of dye dilution with thermodilution data on cardiac output: plot of the difference between the two determinations against their average value. The mean difference (dye minus thermo) is $-2{\cdot}11\,l\,min^{-1}$ (middle horizontal line). The upper and lower horizontal line mark two standard deviations of this mean.

Whereas the dye dilution technique gave an average of $4{\cdot}68 \pm 1{\cdot}42\,l\,min^{-1}$, the corresponding figure by thermodilution was $6{\cdot}80 \pm 1{\cdot}09\,l\,min^{-1}$. A plot according to Bland and Altman[5] reveals the systematic difference, the standard deviation of which appeared to be $1{\cdot}76\,l\,min^{-1}$ (Fig. 3). In other words, the thermodilution technique may overestimate cardiac output by as much as $5\,l\,min^{-1}$. Russell and associates found a mean difference between thermodilution and dye dilution data of $1{\cdot}85\,l\,min^{-1}$ with a standard deviation of the difference of $1{\cdot}24\,l\,min^{-1}$ [4]; thus their figures correspond very well with ours. In addition, they observed a greater difference between the two methods at higher cardiac outputs.

Clinical applications

Despite its cumbersome set-up the dye dilution technique is a useful aid for the experienced investigator wishing to evaluate the haemodynamic status during rest, during exercise or during various types of treatment[10]. With this method, one can easily demonstrate that cardiac output decreases with age[2,11], a phenomenon that may not be discovered by other techniques[12]. Irrespective of age, output at rest also falls with increasing levels of mean arterial pressure[13]. However, there are circumstances in which changes in cardiac output are difficult to interpret. For instance, in an early study from our laboratory it was found that treatment with propranolol by and large prevented the rise in cardiac output elicited by hyperosmotic saline infusion[14]. At rest, output fell from $5{\cdot}5 \pm 1{\cdot}4\,l\,min^{-1}$ to $4{\cdot}7 \pm 1{\cdot}2\,l\,min^{-1}$. Even though this represents an expected reduction, the data from Fig. 2 illustrate that this may still be methodological rather than pharmacological in origin. Even the use of a placebo control group would not entirely settle this issue. Repeated measurements over a prolonged period of time, therefore, are necessary for proper interpretation of data.

Conclusions

The dye dilution technique still is considered the standard among methods available to determine cardiac output. One should bear in mind that there are many potential errors involved in the measurements that may limit its accuracy. These include varying cardiac output during the period that blood is sampled, the rate of withdrawal of blood, the appearance time of the dye, the completeness (or rather the lack thereof) of mixing of the dye with blood, and variations of the volume of injected dye. Although some of these factors may be standardized, instability of cardiac output during the withdrawal period can be neither eliminated nor detected by this averaging technique. It may not be surprising, therefore, that duplicate determinations can vary widely. On balance, however, reproducibility results are not so bad. Moreover, it is striking that a uniform finding is the tendency to record lower values for cardiac output when this variable is repeatedly measured over a short period of time in the same session. Presumably, this reflects the dissipation of the 'alarm reaction' inherent to the procedure. It also suggests that one should obtain multiple determinations for an estimate of 'true' cardiac output; it may often, if not always, be necessary to discard the first measurement.

Differences between estimates of cardiac output by two methods likewise may arise both from errors associated with the techniques themselves and from biological factors. For any comparison between different methods, therefore, measurements must be taken at exactly the same time. Unfortunately, studies in the literature do not always emphasize this point. Among the invasive methods available at present, the thermodilution technique is most often used in clinical medicine. Compared with the standard technique, however, this approach is likely to overestimate cardiac output substantially[4,15], which may be due partly to loss of indicator before and after the injection[16,17]. Whereas thermodilution measurements may be helpful to study sequential changes within an individual, dye dilution data are preferred when absolute data are required. An additional advantage of the dye method is that it can reliably be applied during exercise.

References

[1] Lund-Johansen P. The dye dilution method for measurement of cardiac output. Eur Heart J 1990; 11(Suppl I): 6–12.

[2] Brandfonbrener M, Landowne M, Shock NW. Changes in cardiac output with age. Circulation 1955; 12: 557–66.

[3] Sannerstedt R. Hemodynamic response to exercise in patients with arterial hypertension. Acta Med Scand 1966; 458(Suppl): 7–83.

[4] Russell AE, Smith SA, West MJ *et al.* Automated non-invasive measurement of cardiac output by the carbon dioxide rebreathing method: comparisons with dye dilution and thermodilution. Br Heart J 1990; 63: 195–9.

[5] Bland MJ, Altman DG. Statistical methods for assessing agreement between two methods of clinical measurement. Lancet 1986; i: 307–10.

[6] White SW, Quail AW, de Leeuw PW *et al.* Impedance cardiography for cardiac output measurement: An evaluation of accuracy and limitations. Eur Heart J 1990; 11(Suppl I): 79–92.

[7] Alpert BS, Bloom KR, Gilday D, Olley PM. The comparison between non-invasive and invasive methods of stroke volume determination in children. Am Heart J 1979; 98: 763–6.

[8] Weisel RD, Berger RL, Hechtman HB. Measurement of cardiac output by thermodilution. N Engl J Med 1975; 292: 682–4.

[9] Moodie DS, Feldt RH, Kaye MP, Danielson GK, Pluth J, O'Fallon M. Measurement of postoperative cardiac output by thermodilution in pediatric and adult patients. J Thorac Cardiovasc Surg 1979; 78: 796–8.

[10] Lund-Johansen P, Omvik P. Hemodynamic patterns of untreated hypertensive disease. In: Laragh JH, Brenner BM, eds. Hypertension: pathophysiology, diagnosis and management. New York: Raven Press Ltd; 1990: 305–27.

[11] De Leeuw PW, Kho TL, Falke HE, Birkenhäger WH, Wester A. Haemodynamic and endocrinological profile of essential hypertension. Acta Med Scand 1978; 622(Suppl): 9–86.

[12] Fleg JL. Alterations in cardiovascular structure and function with advancing age. Am J Cardiol 1986; 57: 33C–43C.

[13] Birkenhäger WH, De Leeuw PW. Cardiac aspects of essential hypertension. J Hypertension 1984; 2: 121–5.

[14] Krauss XH, Schalekamp MAD H, Kolsters G, Zaal GA, Birkenhäger WH. Effects of chronic beta-adrenergic blockade on systemic and renal haemodynamic responses to hyperosmotic saline in hypertensive patients. Clin Sci 1972; 43: 385–91.

[15] Van Grondelle A, Ditchey RV, Groves BM, Wagner WW, Reeves JT. Thermodilution method overestimates low cardiac output in humans. Am J Physiol 1983; 245: H690–2.

[16] Mackenzie JD, Haites NE, Rawles JM. Method of assessing the reproducibility of blood flow measurement: factors influencing the performance of thermodilution cardiac output computers. Br heart J 1986; 55: 14–24.

[17] Nadeau S, Noble WH. Limitations of cardiac output measurements by thermodilution. Can Anaesth Soc J 1986; 33: 780–4.

Thermodilution method for measuring cardiac output

J. CONWAY AND P. LUND-JOHANSEN*

*Department of Cardiovascular Medicine, John Radcliffe Hospital, Headington, Oxford, U.K. and *Section of Cardiology, University of Bergen, Haukeland Hospital, 5021 Bergen, Norway*

KEY WORDS: Indicator dilution, accuracy, reproducibility.

The thermodilution technique has been shown to measure cardiac output accurately. The technique violates ideal conditions for indicator dilution methods and is liable to gross errors unless certain requirements are strictly adhered to. The quantity of injectate must be accurately measured and injected rapidly. The dilution curve must be available for inspection and show a smooth rapid ascent and monoexponential decline.

Introduction

Thermodilution, like dye dilution, has been extensively used to measure cardiac output using the following simplified formula:

$$\text{Cardiac output} = \frac{V_I \times (T_B - T_I) \times 60 \times 1{\cdot}08}{A\,(l\,mm^{-1})}$$

where V_I = volume of injectate in ml, T_B = blood temperature, T_I = injectate temperature, 1·08 = correction factor specific gravity and specific heat of blood and indicator, A = the area under the dilution curve multiplied by time for inscription of the curve.

The requirements for an indicator dilution technique to estimate accurately blood flow are: (1) the amount of indicator injected must be measured accurately and there should be no loss of indicator between the sites of injection and detection; (2) the injection should be 'instantaneous'; (3) there must be adequate mixing of the injectate with the volume flow and there must be no additional blood leaving or entering the circulation beyond the site of mixing and before the site of detection; (4) steady-state conditions must apply.

It is obvious that these conditions do not exist in the circulation and all estimates of cardiac output therefore are subject to some error. It is fortunate that none of the deviations is sufficiently severe to prohibit the use of indicator dilution techniques for the measurement of cardiac output[1], but care has to be taken to limit the deviations from the ideal and attention to detail is essential for accurate estimations.

Address for correspondence: James Conway, Department of Cardiovascular Medicine, John Radcliffe Hospital, Headington, Oxford OX3 9DU, U.K.

The dye-dilution technique is described by Lund-Johansen in this issue. The technique is well established and measures cardiac output with reasonable accuracy. Thermodilution, which was introduced by Fegler[2,3] has also found its place and it is more convenient to use than the dye dilution technique because:

(1) there are accurate and rapidly responding thermistors which have a linear output;
(2) proper operation of the system can be checked by calibration of the thermistors using fixed resistors;
(3) the detection system is entirely internal, but a steady-state baseline blood temperature should be present;
(4) frequent serial measurements can be made;
(5) the technique requires only one invasive procedure.

Nevertheless, the thermodilution technique has its particular problems, the greatest of which is accurate assessment of the amount of indicator injected.

Injectate

Saline or glucose are almost universally used as indicators for this technique. Accurate assessment of the amount of injectate is particularly important since it is the denominator in the thermodilution equation. Furthermore, the greater the size of the injectate relative to flow the greater the measurement accuracy, but there has to be a compromise between the desire to inject large volumes and the possible effect that the injectate would then have on flow itself. With large volumes it is also difficult to inject rapidly: full injection should be achieved

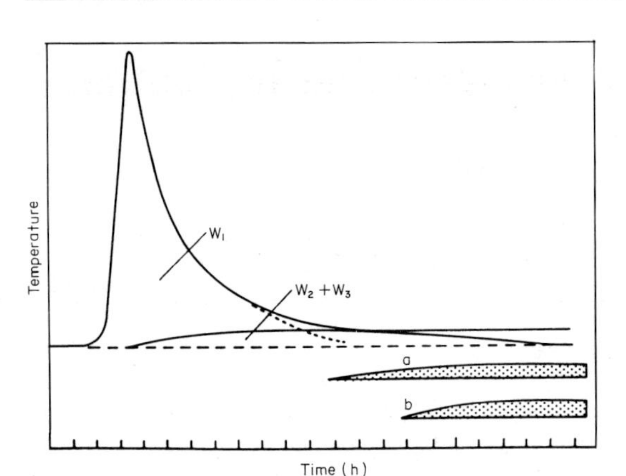

Figure 1 Schematic diagram of the components of a thermal dilution curve recorded from the pulmonary artery after the injection of indicator into the right atrium.
W_1 indicates the curve produced by the indicator actually injected through the catheter.
W_2+W_3 represent the area of the curve produced by conduction of heat through the catheter wall during and after injection of indicator.
a+b represent areas of the curve produced by recirculation of indicator from the superior and inferior vena cava respectively.
Note that any delay in the injection or a low cardiac output will lengthen the primary curve and may cause the areas $W_{2,3}$ and a+b to be included in the area (W) of the dilution curve. This would lead to an underestimate of flow.
Reproduced with permission from Vliers, Visser and Zylstra, *Cardiovascular Research* 1973; 7: 125–32.

within about 2 s. On balance, 10 ml of saline has been found to be reasonable for adults, although 5 ml may also be used. For the larger volume, a powered injector is essential and it is also desirable when 5 ml is used. To increase the 'dose' of injectate, saline can be cooled to ice temperature, and this is commonly done, since the colder the injectate the smaller the variability of the cardiac output estimate[4]. However, using cold injectate requires special care to ensure the temperature does not rise as the cold saline is drawn up into a syringe and injected. It is preferable to use a catheter with an additional thermistor lying in the right atrium about 8 cm from the tip of the catheter, which can measure the injectate temperature at the point of delivery, in the right atrium.[5]

Obviously, the volume in the syringe has to be measured accurately and, from this, the volume of injectate remaining within the catheter after injection must be subtracted. A further problem in estimating the precise 'dose' is the passage of heat from the blood to the saline within the catheter after injection. With cold saline, this can cause an underestimation of cardiac output by 2–5%, depending on the cardiac output[6,7]. This error can be largely overcome by withdrawal of saline from the catheter after injection[5].

There is inevitably a loss of injectate temperature due to heat passing from the walls of the heart into the blood as the injectate passes to the sampling site. It was first thought[1] that this would invalidate the method, but the problem is not severe because after the injectate has passed to the detector heat will be returned to the tissues. This results only in a small distortion of the indicator dilution curve (Fig. 1).

Detection

It has been established that there is adequate mixing of the saline between the right atrium and the pulmonary artery[8]. Accordingly, the detection site in the pulmonary artery is satisfactory, and it has three advantages: there is a smaller loss of indicator to the tissues than if the cool blood had to pass through the lungs before detection[9]; the duration of the 'tail' of the indicator dilution curve is reduced; and not only is the duration of the dilution curve shorter, but also the temperature change seen by the detector is greater. These factors greatly improve the accuracy of the method[7,9].

A disadvantage of the pulmonary artery as a detector site is that there are respiratory fluctuations in the temperature of blood in the pulmonary artery. This is due to different temperatures of the blood returning from different areas of the body. Blood from these several sites enters the thorax with inspiration. Respiratory fluctuation in temperature leads to some inevitable 'noise' in the system, and this is one of the main reasons for the recommendation of a large volume of injectate, 10 ml, and very cold saline. The temperature fluctuation with respiration is approximately 0·05 °C, which is minor since the mean change in temperature is of the order of 0·5 °C. Changes in physiological condition may also induce variations in temperature. This has been clearly shown, for example, after injection of adrenaline, but can also occur with exercise and, of course, laboured breathing[9]. The change in temperature recorded at the pulmonary site after the injection of cold saline should rise to a maximum quickly and then decline with monoexponential kinetics. Close to baseline an extra tail will be present due to 'loss' of indicator into the walls of the ventricles, as mentioned above (Fig. 1). If computers are used to measure the area of the thermodilution curve it is desirable that they integrate the entire curve down to the point, close to the baseline, where the curve deviates from mono-exponential decay, and then extrapolate the rest of the curve. Computers must also display the indicator dilution curve to enable the operator to verify a rapid and smooth build-up of the curve and an exponential decay towards baseline[10]. Comparisons of computers in a model circulation has revealed substantial differences between their performance[4].

The response time of thermistors is adequate to follow these curves but, when the catheters have been in position for some time, fibrin may accumulate on the surface of the thermistor and damp the indicator dilution curve. The frequency response of the thermistors is sufficient to demonstrate the presence of respiratory fluctuations and include them in the area estimated.

The greatest value of the thermodilution technique is that curves can be repeated several times without disturbing the baseline and two curves can be done a minute apart. This is extremely valuable for a method of this kind with its susceptibility to error. Repeat curves should agree with one another to within 7%. Estimations should be made at different phases of respiration and averaged.

Limitations of the procedure

Thermodilution cannot be used in the presence of circulatory shunts or significant disease of the tricuspid or pulmonary valves. The technique tends to become inaccurate at very low flow rates when recirculation can occur before the primary curve has been fully described. This is indicated by deviation from a mono-exponential decay curve. Laboured breathing or exercise can lead to instability of baseline blood temperature. Acute changes in cardiovascular status can, of course, invalidate the method.

Comparison with standard methods

THERMODILUTION VS DIRECT MEASUREMENTS OF FLOW

The thermodilution method has shown no systematic differences in estimation of flow compared with absolute measurement (timed collection of fluid) in models of the circulation[3,4,10,11], the mean difference between the two methods being about 4–5%[4,10]. The reproducibility is also satisfactory, about 0·5 l min^{-1} at flow rates of 5 l mm^{-1}. Models of the circulation have also provided the opportunity to compare the performance of three computer systems. These were far from satisfactory, since confidence limits varied from 30–62% of actual flow[4]. Inaccuracies were greater at high rates of flow.

THERMODILUTION AND DYE DILUTION

Although one report has shown that thermodilution tended to give a cardiac output value 5% greater than the dye method[8], others have shown no such trend[10–12]. The correlation coefficient between the two methods is high (0·91–0·98)[5,11], the ratio between dye and thermal estimates ranges from 1·175 to 0·93[3,12] and the correlation coefficients between the two methods lay between 0·88 and 0·94, with a standard error of 0·36 l mm^{-1} (±7%)[5].

VARIABILITY

Spontaneous changes in cardiac output make it difficult to arrive at an accurate estimate of the variability due to inaccuracies in the method itself. However, when large numbers of measurements are made spontaneous variability becomes less important, and published results are reasonably consistent. The standard deviation between repeat measurements over a short period of time has been given as 1·4–2·2 l min^{-1} at a flow of 5 l mm^{-1} or ±7%[12–14]. Over a wider range of flows (4–16 l mm^{-1}) the standard error of the estimate ranged from 0·71 to 3·9 l min^{-1}[15,16]. When the mean was taken of three measurements of output the standard error varied from 2–5%, which indicates that the method can be used to detect changes of 6–15%[17,18]. Data on longer term variability are not available.

Conclusion

The thermodilution method for measuring cardiac output has been shown to be satisfactory providing fairly strict conditions are met. The method violates several conditions required for accuracy and it is essential that certain criteria are strictly adhered to.

A rapid injection of 5 ml, or preferably 10 ml, of ice-cold saline into the right atrium, and detection in the pulmonary artery are the most reliable conditions. Mechanically powered injection is essential for these volumes. The quantity and temperature of the cold saline must be measured accurately and, although some loss of indicator is unavoidable, the error can be minimized if the cold saline remaining in the catheter is withdrawn after injection.

Modern thermistors are easily calibrated and have an adequate response time, but build-up of fibrin can produce damping of the response. Cardiac output computers, to be satisfactory, must display the dilution curve and should record and integrate the whole of the indicator dilution curve to the point of deviation from the mono-exponential decay and then extrapolate to baseline. The method is particularly vulnerable to error at low cardiac output.

References

[1] Dow P. Estimations of cardiac output and central blood volume by dye dilution. Physiol Rev 1956; 36: 77–102.
[2] Fegler G. Measurement of cardiac output in anesthetized animals by a thermodilution method. Quart J Exptl Physiol 1954; 39: 153–64.
[3] Fegler G. The reliability of the thermodilution method for determination of the cardiac output and the blood flow in central veins. Quart J Exptl Physiol 1957; 42: 254–66.
[4] Mackenzie JD, Haites NA, Rawles JM. Method of assessing the reproducibility of blood flow measurement: factors influencing the performance of thermodilution cardiac output computers. Br Heart J 1986; 55: 14–24.
[5] Ganz W, Donoso R, Marcus HS, Forrester JS, Swan HJC. A new technique for measurement of cardiac output by thermodilution in man. Am J Cardiol 1971; 17: 392–6.
[6] Ganz W, Swan HJC. Measurement of blood flow by thermodilution. Am J Cardiol 1972; 29: 241–6.
[7] Vliers ACAP, Visser KR, Zulstra WG. Analysis of indicator distribution in the determination of cardiac output by thermal dilution. Cardiovasc Res 1973; 7: 125–32.
[8] Vliers ACAP, Oesburg B, Visser KR, Zijlstra WG. Choice of detection site for the determination of cardiac output by thermal dilution: the injection-thermistor catheter. Cardiovasc Res 1973; 7: 133–8.
[9] Wessel HU, Paul MH, James GW, Grahn AR. Limitations of thermal dilution curves for cardiac output determinations. J Appl Physiol 1971; 30: 5.
[10] Bilfinger TV, Lin CY, Anagnostropoulos CE. In vitro determination of accuracy of cardiac output measurements by thermal dilution. J Surg Res 1982; 33: 409–14.
[11] Runciman WB, Isley AH, Roberts JG. Thermodilution cardiac output: A systematic error. Anaes Intens Care 1981; 9: 135–9.
[12] Fischer AP, Benis AM, Jurado RA, Seely E, Teirstein P, Litwak RS. Analysis of errors in measurement of cardiac output by simultaneous dye and thermal dilution in cardiothoracic surgical patients. Cardiovasc Res 1978; 12: 190–9.
[13] Sorensen MB, Bille-Brake NE, Engell HC. Cardiac output measurement by thermal dilution. Reproducibility and comparison with dye-dilution. Ann Surgery 1976; 183: 67–72.
[14] Sanmarco ME, Philips CM, Marquez LA, Hall C, Davila JC. Measurement of cardiac output by thermal dilution. Am J Cardiol 1971; 28: 54–8.
[15] Shellock FG, Riedinger MS. Reproducibility and accuracy of using room-temperature vs ice-temperature injectate for thermodilution cardiac output determination. Heart Lung 1983; 12: 175–6.
[16] Ollson B, Pool J, Vandermoten P, Varnausks E, Wassen R. Validity and reproducibility of determination of cardiac output by thermodilution. Cardiology 1970; 55: 136–48.
[17] Stetz WC, Miller RG, Kelly GE, Raffin TA. Reliability of the thermodilution method of determining cardiac output in clinical practice. Am Rev Respir Dis 1982; 26: 1001–4.
[18] Kadota LT. Theory and application of thermodilution cardiac output measurement: A review. Heart Lung 1985; 14: 605–16.

Assessment of cardiac output at rest and during exercise by a carbon dioxide rebreathing method

T. Reybrouck and R. Fagard

Hypertension and Cardiovascular Rehabilitation Unit, University Hospitals, University of Leuven (K.U. Leuven), Belgium

KEY WORDS: CO_2 rebreathing method, exercise.

The validity of a CO_2 rebreathing (exponential) method to determine cardiac output was acceptable during exercise, but not at rest. The reproducibility during exercise of the measurement of cardiac output was similar to data published for the dye dilution method.

Introduction

The original Fick method, when modified to use CO_2 as the indicator, allows assessment of cardiac output non-invasively; CO_2 output per minute is measured and the arteriovenous difference in CO_2 content is estimated from alveolar gas concentrations. The arterial CO_2 content can be estimated from end tidal pCO_2, and mixed venous pCO_2 can be derived from measurements taken during rapid rebreathing in an anaesthesia bag.

Two methods are currently used for the estimation of mixed venous pCO_2, which differ in the initial CO_2 concentration (FCO_2) in the rebreathing bag. In the 'equilibration' or Collier procedure, the FCO_2 in the bag is higher than the mixed venous FCO_2. If an appropriate initial FCO_2 and bag volume have been chosen, the gas in the bag will mix with alveolar FCO_2, generating an equilibrium between the lung bag system and the mixed venous pCO_2[1]. In the exponential method, FCO_2 in the bag is lower than mixed venous FCO_2. The mixed venous pCO_2 is derived mathematically from the asymptotic breath-by-breath increase in end-tidal pCO_2 during rebreathing[2–4].

The purpose of the present study was to establish the validity of one CO_2 rebreathing method (the exponential method) to measure cardiac output at rest and during exercise.

Methods

In 16 hypertensive patients, 59 cardiac output estimates were made by the CO_2 rebreathing method and direct Fick method for O_2, at rest and during exercise, on the occasion of a haemodynamic investigation. The patients were selected on the basis of normal lung function tests. The clinical characteristics and pulmonary function data for these patients have been published previously[3]. Their mean age was 35 ± 10 (SD) years. Some patients were receiving beta-blocking agents. To study the reproducibility of cardiac output determination by the CO_2 rebreathing method, duplicate measurements of cardiac output were compared with those from six healthy volunteers who participated in another study[5]. The mean age of the latter subjects was $29 \pm 7{\cdot}2$ years. All subjects agreed to the experimental procedure after the nature of the test had been fully explained.

In the hypertensive patients, cardiac output was estimated in two ways: first according to the direct Fick principle for O_2 and for CO_2:

$$\dot{Q} = \dot{V}O_2/(Ca,O_2 - C\bar{v},O_2) \qquad [1]$$

or

$$\dot{Q} = \dot{V}CO_2/(C\bar{v},CO_2 - Ca,CO_2) \qquad [2]$$

where $\dot{Q}$ is cardiac output expressed in $l\,min^{-1}$, $\dot{V}O_2 = O_2$ uptake in $ml\,min^{-1}$, $\dot{V}CO_2 = CO_2$ output in $ml\,min^{-1}$, $C\bar{v},O_2$ and $C\bar{v},CO_2 =$ ml of O_2 or $CO_2\,l^{-1}$ of mixed venous blood, and Ca,O_2 or $Ca,CO_2 =$ ml of O_2 or $CO_2\,l^{-1}$ of arterial blood. Right-heart catheterization was performed, with a Swan-Ganz catheter floated into the pulmonary artery. The brachial artery was catheterized with a Teflon needle. Blood samples were drawn slowly and simultaneously from the brachial and pulmonary arteries. The oxygen content of arterial and mixed venous blood was calculated from the haemoglobin concentration (cyanmethaemoglobin method), the

Address for correspondence: Dr T. Reybrouck, Department of Cardiovascular Rehabilitation, University Hospital, Weligerveld 1, 3212 Pellenberg, Belgium.

oxygen saturation (Sa,O_2) and combining factor 1·34. The Sa,O_2 was analysed by a reflection oximeter.

pCO_2 was determined in whole blood by a Teflon-covered glass electrode. Before each measurement, the pCO_2 electrode was calibrated against a gas mixture of known composition, which had previously been analysed by gas chromatography. The CO_2 content was derived from a standard dissociation curve, by taking into account the Sa,O_2 of the mixed venous blood[6].

The subjects breathed through a low-resistance Y-valve. The pulmonary ventilation ($\dot{V}_E$) was measured by a pneumotachograph, and oxygen uptake ($\dot{V}O_2$) and CO_2 output ($\dot{V}CO_2$) were continuously calculated from the volume, O_2 and CO_2 composition of expired air collected in a 5 l mixing box. Gas concentrations were analysed by a paramagnetic O_2 analyser and an infrared CO_2 analyser, both previously calibrated with test gases of known composition. $\dot{V}O_2$ and $\dot{V}CO_2$ values were reduced to STPD. Each variable was printed out every 12 s on a polygraph. All experiments were performed in an air-conditioned laboratory at 18–22 °C and humidity 40–60%.

Second, cardiac output was also determined by the indirect Fick method, where mixed venous pCO_2 ($P\bar{v},CO_2$) was estimated by the rebreathing method of Defares[2,7,8]. The $P\bar{v},CO_2$ was extrapolated from the exponential increase in CO_2 during rebreathing, as described by Klausen[9]. The 4 l rebreathing bag was filled with sufficient gas (4% CO_2 and 96% O_2) to equal the actual tidal volume. The rebreathing rate was 30 breaths min^{-1} and rebreathing was usually complete within 12 s. Arterial pCO_2 (Pa,CO_2) was estimated from the average value of at least 10 end-tidal pCO_2 readings preceding the rebreathing manoeuvre. To convert arterial and mixed venous pCO_2 into CO_2 content a standard CO_2 dissociation curve was used for whole oxygenated blood[6].

At rest, in the recumbent and sitting positions, and during graded exercise, 59 comparisons were made between the direct Fick method and CO_2-rebreathing estimates of cardiac output. Exercise was performed on a bicycle ergometer (Elema Schönander) and exercise loads varied from 20 to 110 W. In normal volunteers, the same exercise procedure was used and exercise intensities varied from 20 to 170 W.

Table 1. Respiratory exchange ratio as measured in the blood and respired gas

State	Respiratory exchange ratio		n
	Blood	Respired gas	
Rest, recumbent	0·89±0·16	0·83±0·05	13
Rest, sitting	0·76±0·09	0·82±0·04**	16
Work at 20 W	0·98±0·25	1·02±0·26	6
Work at 50 W	0·91±0·13	0·88±0·10	17
Work at 80 W	1·00±0·15	0·99±0·14	7
Work at 110 W	1·11±0·14	1·07±0·10	5

Statistical significance refers to a paired comparison (** = $P<0·01$). Values represent the mean ±SD of the mean. Reprinted with permission from Reybrouck *et al.*[3]

STATISTICAL ANALYSIS

The relationship between cardiac output estimated by the CO_2 rebreathing method and by the direct Fick method for O_2 was initially assessed from the correlation coefficient. Agreement between both methods was also assessed by calculating the difference between the values of cardiac output estimated by the two methods and plotting this difference against their mean value, as described by Bland & Altman[10]. Reproducibility was taken as the percentage difference between two measurements, calculated as 100 times absolute difference divided by the mean value of the true measurement.

Results

When the blood and gas respiratory gas exchange ratios were compared, no significant difference was found between mean values, except at rest in the sitting position (Table 1). However, when blood R was plotted against respiratory gas R, 33% of the values were more than 10% away from the line of identity.

VALIDITY OF CARDIAC OUTPUT DETERMINATION BY CO_2 REBREATHING

When cardiac output estimates by the CO_2 rebreathing method were compared with those by the direct Fick method, a good correlation ($r=0·96$, $P<0·001$, $n=59$) between the methods was found. Only 12% of the exercise cardiac output values determined consecutively by the direct Fick method for O_2 and CO_2 rebreathing were more than 10% from the line of identity, vs 48% of cardiac output values at rest.[3]

In Fig. 1, the difference between the two methods is plotted against the mean value for cardiac output, determined at rest. The limits of agreement (= ±2

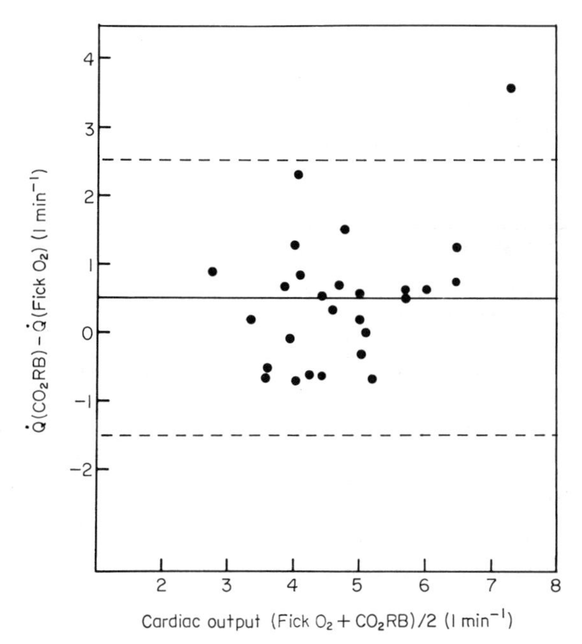

Figure 1 Difference against mean value of cardiac output ($\dot{Q}$) at rest, determined by the CO_2 rebreathing method and the direct Fick method for O_2.

SD above and below the mean value) were $+2{\cdot}5$ and $-1{\cdot}5\,l\,min^{-1}$. The cardiac output was overestimated by the CO_2 rebreathing method by an average $0{\cdot}5\,l\,min^{-1}$, a statistically significant difference ($P<0{\cdot}05$). During exercise, on the other hand, the mean difference was negligible ($-0{\cdot}16\,l\,min^{-1}$; $P>0{\cdot}25$) and the limits of agreement were less than at rest: varying from $+1{\cdot}3$ to $-1{\cdot}6\,l\,min^{-1}$ (Fig. 2).

REPRODUCIBILITY OF CARDIAC OUTPUT MEASUREMENTS BY CO_2 REBREATHING

The percentage differences between duplicate values for cardiac output, in six healthy male volunteers at rest and during exercise are given in Table 2. The highest values were found at rest ($24{\cdot}5 \pm 23\%$). During exercise, however, the percentage difference between duplicate estimates of cardiac output at the same exercise intensity decreased and reached values of $8 \pm 6{\cdot}6\%$ at 110 W.

Discussion

During exercise, no significant difference was found between cardiac output values determined by the CO_2 rebreathing method and the direct Fick method for O_2. This suggests that no systematic error occurs when cardiac output is estimated during exercise by the CO_2 rebreathing method. Also the limits of agreement between both methods were smaller during exercise than at rest. Non-systematic errors may result from technical differences and errors inherent in both methods, and also intra-individual variation, since the cardiac output estimate by CO_2 rebreathing always followed the cardiac output determination by the direct Fick method for O_2. The difference in R values ($> \pm 10\%$) in about one-third of the patients suggests that a failure to reach steady-state could contribute to some difference in cardiac output estimates. It has been shown by Sietsema *et al.*[11] that the time to reach one time constant or 63% of the steady-state value for $\dot{V}O_2$ was higher in patients with impaired cardiovascular function than in normal controls. The maximal exercise capacity of the patients in the present study was rather low[3].

At rest, however, the validity of the CO_2 rebreathing method (exponential method) is questionable; the limits of agreement between the rebreathing and direct Fick method were almost twice that during exercise. This is in agreement with other studies[12,13] comparing the validity of the CO_2 rebreathing method to determine cardiac output with that of the dye dilution method. Although, during exercise, mixed venous pCO_2 was not significantly different when measured by exponential and equilibrium methods in the study of Heigenhauser and Jones[4], it has been reported that, at rest, the direct Fick or dye dilution method shows closer agreement with the equilibrium method than the exponential method[14,15]. It is possible that the equilibrium method is more appropriate to determine cardiac output at rest than is the extrapolation technique. To eliminate bias inherent in the graphical extrapolation procedure of $P\bar{v}CO_2$, as described by Klausen[9], mathematical procedures using iterative techniques to estimate $P\bar{v}CO_2$, have been shown to provide valid estimates of cardiac output[16].

The percentage difference between duplicate determination of cardiac output was largest at rest and decreased during exercise. The exercise values (8–19%) were comparable to those calculated from data reported by Grimby *et al.*[17] for the dye dilution method (9–17%), who made serial measurements of cardiac output every 2–5 min during a 30-min exercise period (Table 2).

It should be emphasized that for cardiac output determinations by the indirect Fick method for CO_2, all elements are not determined simultaneously. The

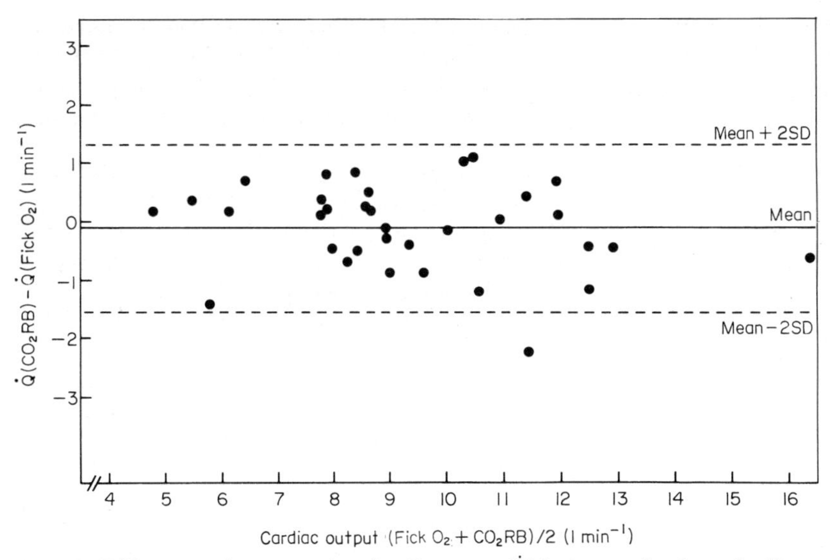

Figure 2 Difference against mean value of cardiac output ($\dot{Q}$) during exercise, determined by the CO_2 rebreathing method and the direct Fick method for O_2.

Table 2. Duplicate determinations of cardiac output (see text)

Present study (CO_2 rebreathing method)		Grimby *et al.*[17] (dye dilution)	
*percentage difference**			*percentage difference*
– at rest, recumbent	22·3 ± 18·8%		
– at rest, sitting	24·5 ± 23%		
– exercise: 50 W	19·9 ± 15·6%		
– exercise: 110 W	8 ± 6·6%	(100 W)	2–5 min: 17·5 ± 9·9%
			5–7 min: 8·8 ± 3·7%
– exercise: 170 W	12·9 ± 7·3%	(150 W)	2–5 min: 16 ± 11·2%
			5–7 min: 11·7 ± 12%

Values are means ± SD. *Absolute difference × 100/average of two measurements.

determination of $P\dot{V}O_2$ always followed the estimation of end-tidal pCO_2 and $\dot{V}CO_2$. With exercise intensities above the anaerobic threshold (i.e. with accumulation of lactate) a steady state for $\dot{V}O_2$ and $\dot{V}CO_2$ between the 3rd and 6th min of exercise is often lost[18], which complicates the study of reproducibility of cardiac output.

If we consider ⩾10% variation of cardiac output acceptable for repeated measurement in the same subject[19], the reproducibility of the CO_2 rebreathing method to determine cardiac output during exercise is similar to that of invasive techniques[4,12,20].

In conclusion, cardiac output can be measured non-invasively by applying the Fick principle for CO_2. Its accuracy during exercise is acceptable, whereas at rest it is less reliable.

The authors gratefully acknowledge the secretarial assistance of Mrs S. Teck.

References

[1] Jones NL, Campbell EJM. Clinical exercise testing. 2nd edn. Philadelphia: WB Saunders, 1981.
[2] Defares JG. Determination of $P\bar{v}CO_2$ from the exponential CO_2 rise during rebreathing. J Appl Physiol 1958; 13: 159–64.
[3] Reybrouck T, Amery A, Billiet L, Fagard R, Stijns H. Comparison of cardiac output determined by a carbon

dioxide-rebreathing and direct Fick method at rest and during exercise. Clin Sci Mol Med 1978; 55: 445–52.
[4] Heigenhauser GJF, Jones NL. Measurement of cardiac output by carbon dioxide rebreathing methods. Clin Chest Med 1989; 10: 255–64.
[5] Fagard R, Amery A, Reybrouck T *et al.* Effects of angiotensin antagonism on hemodynamics, renin and catecholamines during exercise. J Appl Physiol 1977; 43: 440–7.
[6] Comroe JH, Forster RE, Dubois AG, Briscoe WA, Carlsten E. The lung. Clinical physiology and pulmonary function tests. Chicago. Year Book, 1957.
[7] Jernerus R, Lundin G, Thompson D. Cardiac output in healthy subjects determined with a CO_2 rebreathing method. Acta Physiol Scand 1963; 59: 390–9.
[8] Heigenhauser GJF, Jones NL. Comparison of two rebreathing methods for the determination of mixed venous partial pressure of carbon dioxide during exercise. Clin Sci 1979; 56: 433–7.
[9] Klausen K. Comparison of CO_2 rebreathing and acetylene methods for cardiac output. J Appl Physiol 1965; 20: 763–6.
[10] Bland JM, Altman DG. Statistical methods for assessing agreement between two methods of clinical measurement. Lancet 1986; i: 307–10.
[11] Sietsema KE, Cooper DM, Perloff JK *et al.* Dynamics of oxygen uptake during exercise in adults with cyanotic congenital heart disease. Circulation 1986; 73: 1137–44.
[12] Ferguson RJ, Faulkner JA, Julius S, Conway J. Comparison of cardiac output determined by CO_2 rebreathing and dye dilution methods. J Appl Physiol 1968; 25: 450–4.
[13] Clausen JP, Larsen OA, Trap-Jensen J. Cardiac output in middle-aged patients determined with CO_2 rebreathing method. J Appl Physiol 1970; 28: 337–42.
[14] Muiesan G, Sorbini CA, Solinas E, Grassi V, Casucoi G, Petz E. Comparison of CO_2 rebreathing and direct Fick methods for determining cardiac output. J Appl Physiol 1968; 24: 424–9.
[15] Hinderlinten AL, Kitzpatrick A, Schork N, Julius S. Research utility of noninvasive methods for measurement of cardiac output. Clin Pharmacol Ther 1987; 41: 419–25.
[16] McKelvie RS, Heigenhausen GJF, Jones NL. Measurement of cardiac output by CO_2 rebreathing in unsteady state exercise. Chest 1987; 92: 777–82.
[17] Grimby G, Nilsson NJ, Sanne H. Repeated serial determination of cardiac output during 30 min exercise. J Appl Physiol 1966; 21: 1750–6.
[18] Wasserman K. Determinants and detection of anaerobic threshold and consequence of exercise above it. Circulation 1987; 76 (Suppl VI): 29–39.
[19] Driscoll DJ, Staats BA, Beck KC. Measurement of cardiac output in children during exercise: a review. Ped Ex Sc 1989; 1: 102–15.
[20] Ohlsson J, Wranne B. Non-invasive assessment of cardiac output and stroke volume in patients during exercise. Eur J Appl Physiol 1986; 55: 538–44.

Continuous cardiac output monitoring by pulse contour during cardiac surgery

J. R. C. JANSEN, K. H. WESSELING*, J. J. SETTELS* AND J. J. SCHREUDER†

*Pathophysiological Laboratory, Department of Pulmonary Diseases, Erasmus University, Rotterdam, *TNO Biomedical Instrumentation, Academic Medical Centre, Amsterdam, and †Department of Anesthesiology, University of Maastricht, The Netherlands*

KEY WORDS: Continuous cardiac output, thermodilution method, pulse contour method, CABG patients.

Most pulse contour methods are unreliable under changing haemodynamic conditions, because no corrections are made for pressure-dependent compliance and reflections of pressure waves. The pulse contour method of Wesseling includes such corrections. Four thermodilution measurements equally spread over the ventilatory cycle were used to calibrate and evaluate this pulse contour method.

We designed a prototype incorporating a combination of the thermodilution method and pulse contour method and evaluated its potential for monitoring patients undergoing coronary bypass graft operation. Eight to 12 times during the operation, cardiac output was estimated by pulse contour and by thermodilution. The results were compared: the linear regression between the methods was $CO_{pc}=0{\cdot}3+0{\cdot}94 . CO_{th}$, $(r=0{\cdot}94)$. *The standard deviation for the difference between the methods against the mean of the methods was 10·6%. We concluded that the corrected pulse contour method estimates cardiac output accurately, even when heart rate, blood pressure, and total peripheral resistance change substantially.*

Introduction

Continuous monitoring of cardiac output is important for the management of critically ill patients since their condition often changes rapidly and unexpectedly. Until now, no reliable methods have been available to measure cardiac output continuously over a long period.

Unfortunately, the method used most often clinically, the thermodilution technique, allows only intermittent measurements. Moreover, clinically, a combination of three or more thermodilution estimates must be averaged, because individual thermodilution estimates show substantial scatter[1–6].

An attractive method to estimate the beat-to-beat changes in cardiac output is the pulse contour method. It calculates changes in relative stroke volume and in cardiac output from the arterial pressure wave. Most pulse contour methods are unreliable under changing haemodynamic conditions[7–11]. The pulse contour method of Wesseling[12,13] uses a correction factor for changes in blood pressure and heart rate, but still requires calibration against an absolute method, such as thermodilution. A system combining thermodilution and pulse contour is continuous and calibrated and could prove a clinically relevant cardiac output monitor.

We designed a prototype of such a combination of methods and evaluated its potential in patients undergoing coronary bypass graft (CABG) operation, since large changes in haemodynamics occur during cardiac surgery owing to administration of drugs, surgical interventions, cardioplegia, and total cardio-pulmonary bypass.

Correspondence address: Dr J. R. C. Jansen, Dept. of Pulmonary Diseases, Erasmus University, room Ee2251, P.O. Box 1738, 3000 DR Rotterdam, The Netherlands.

Methods

For a clear understanding of the combined thermodilution and pulse contour method the two individual methods will be discussed first.

THE PULSE CONTOUR METHOD

All pulse contour methods are, explicitly or implicitly, based on a model. Most models are haemodynamic in nature, relating an arterial pressure or pressure difference to a flow or volume via the impedance through which the flow is driven. If the systemic circulation is considered to be a

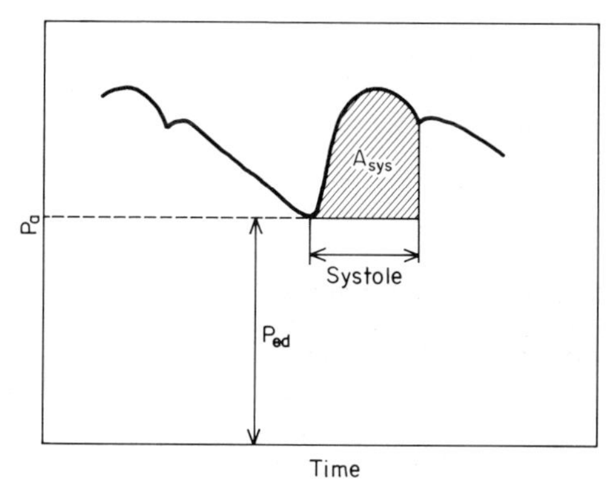

Figure 1 Determination of the individual beat-to-beat 'stroke volume' (SV) from the arterial pressure (P_a). A_{sys} = area under the pressure curve during the left ventricular ejection phase (systole) above the end-diastolic pressure level (P_{ed}).

'Windkessel' model stroke volume can be calculated from the pressure as the driving force for flow during the ejection time (Fig. 1):

$$SV = A_{sys}/Z_{ao} \qquad [1]$$

where SV is the pulse contour stroke volume of the heart, A_{sys} the area under the systolic portion of the pressure wave, Z_{ao} the characteristic impedance of the aorta. Obviously, this model for the human circulation is too simple. However, sufficient information on human arterial haemodynamics is available today to provide an extension on this model[13,14]. Mean arterial pressure is used for the correction of pressure-dependent non-linear changes in cross-sectional area of the aorta, and heart rate is used to correct for reflections from the periphery. (If the time for the pressure wave to travel forwards and backwards is longer than the ejection time no correction is needed.) These corrections for pressure and rate are furthermore age dependent. A detailed description can be found in the paper by Wesseling[13]. Basically, the computation can be written as:

$$CO_{pc} = HR \,.\, A_{sys} \,.\, \frac{1}{Z_{ao}}, \qquad [2]$$

$$Z_{ao} = \frac{a}{b + (c.P_{mean}) + (d.HR)} \qquad [3]$$

where CO_{pc} is the pulse contour stroke volume, HR the heart rate, P_{mean} mean arterial pressure, and a, b, c, and d age-dependent parameters respectively. An initial estimate of Z_{ao} is computed if age, mean pressure and heart rate are available. The actual characteristic impedance Z_{ao} is not known and must be determined at least once for each patient by comparison with an absolute cardiac output estimate:

$$Z_{ao} = (CO_{pc}/CO_{ref}) \,.\, Z_{ao,initial} \qquad [4]$$

For this study we have chosen the thermodilution technique as the reference method CO_{ref}.

THE THERMODILUTION TECHNIQUE

Although the thermodilution method has been evaluated extensively and is accepted by many investigators[15–18], we must remember that the computation is based on the application of the Stewart-Hamilton equation, for which a number of important conditions must be fulfilled. These conditions are: complete mixing of indicator and blood; no loss of indicator between the sites of injection and detection; and constant blood flow.

The errors made in the estimation of cardiac output by thermodilution are related to: (1) violation primarily of the above mentioned condition of constant blood flow. Variability of blood flow occurs during mechanical ventilation, shivering, variations in heart rate, cardiac arrhythmias, valvular insufficiencies, intra-cardiac shunts, and other causes of haemodynamic instability[15,17]; (2) errors due to careless usage of the technique, such as warming of injectate[17,18], incorrect catheter positioning[19], or injections at an uneven rate[17,20]; (3) spontaneous changes in baseline blood temperature in the pulmonary artery[1,4,16,17,19]; (4) cardiac output computer inaccuracy[21]. Hence, the clinician should expect 5–15% data scatter, even in haemodynamically stable patients[21]. There is, however, good evidence in animal studies that the mean of a large number of thermodilution measurements will give an accurate mean cardiac output[1,4,22]. Thus, a high variance indicates the need to increase the number of measurements. The number of consecutive thermodilution determinations that must be averaged per estimation of mean cardiac output depends on the nature of the variance (normal or non-normal distribution). For a normal distribution of errors the standard deviation will decrease with the square root of the number of observations as we indeed observed[4,6].

MEASUREMENT PROTOCOL DURING MECHANICAL VENTILATION

An example of the effects of mechanical ventilation on cardiac output estimates is shown in Fig. 2.

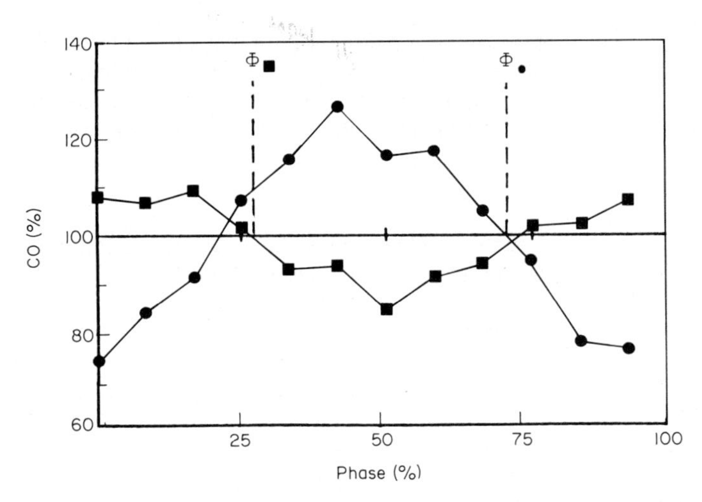

Figure 2 Two individual series of 12 thermodilution cardiac output (CO) measurements, in two patients, plotted against the moment of injection as a percentage of the phase of the ventilatory cycle. 100% CO is mean of each series of 12 estimates. The determinations were done with a time interval of approximately 1 min. Φ indicates the phase in the respective ventilatory cycles at which CO equals 100% of the mean.

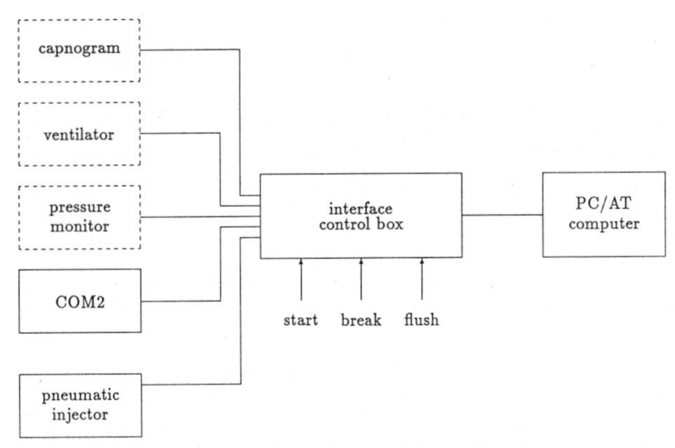

Figure 3 Schematic diagram of the continuous cardiac output monitoring system. The dashed boxes are input signals to the system. For explanation see text.

In a series of 12 measurements equally spread over the ventilatory cycle a cyclic pattern of modulation of the estimates was found with the same periodicity as the ventilation. We observed marked differences in the amplitude and the phase of the pattern of modulation in a group of nine patients[6]. A more detailed analysis of the variation in cardiac output estimates related to the moment of injection in the ventilatory cycle is given elsewhere for animals[1,2,4] and humans[5,6]. In patients, the standard deviation decreased from 14% to 3·1% when four thermodilution measurements equally spread over the ventilatory cycle were averaged, whereas it decreased to 7·2% if four randomly performed measurements were averaged[6]. Therefore, to obtain a reliable estimation of mean cardiac output during mechanical ventilation we have chosen to average four consecutive thermodilution measurements which are performed at four moments equally spread over the ventilatory cycle.

SYSTEM DESCRIPTION

For the continuous cardiac output system a PC/AT computer, an intra-arterial pressure signal,

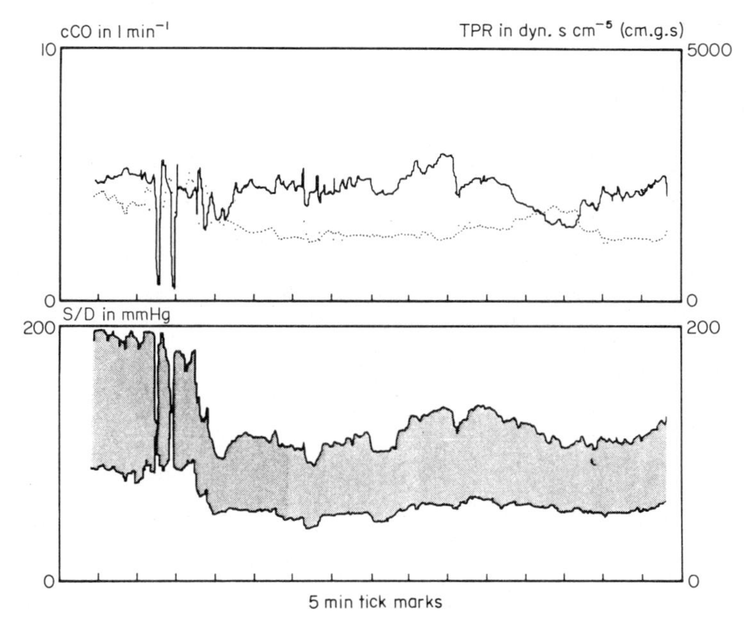

Figure 4 A simplified copy of the screen layout of the continuous cardiac output monitor. cCO, continuous cardiac output by pulse contour (solid line); TPR, total peripheral resistance (dotted line); S/D, systolic and diastolic blood pressure respectively.

a respiratory signal from a ventilator or a capnogram, an Edwards COM-2 thermodilutation cardiac output computer, and a modified Broszeit Medizintechnik injectate pump are used (Fig. 3). All signals to and from the computer are routed via an interface box to provide full patient isolation and to accomplish certain signal conditioning tasks. A red 'break-key' on the front panel of the interface can interrupt automatic injection of cold thermodilution injectate to provide adequate patient protection.

For each beat in the arterial pressure waveform, systolic, diastolic and mean pressure and pulse rate are derived in addition to a measurement of relative left ventricular stroke volume, cardiac output, ejection time and total systemic peripheral resistance. Their values are displayed on a computer screen in various beat-to-beat and averaged forms. At instants determined by the physician, a series of one, two, three or four thermodilution estimates are performed automatically and equally spread over the respiratory cycle, provided a ventilator signal or a capnogram is available. If not, the series is performed randomly. Once a series of four thermodilution estimates has been completed successfully, the average value is used to (re-)calibrate the beat-to-beat cardiac output computed simultaneously from the pressure pulse contour. All data are stored in computer files on hard disc for further offline analysis.

PATIENTS STUDIED

We tested our system in patients undergoing elective CABG operation because this group of patients shows characteristic large and rapid changes in haemodynamics. All patients suffered from multiple vessel diseases, but had normal ventricular function and stable angina pectoris. As premedication, the patients received 5 mg lorazepam 2 h before surgery. Peripheral venous catheters, radial artery cannula (20 G) and 7F Swan-Ganz pulmonary artery catheters were inserted. The radial artery pressure was used as a substitute of the central aortic pressure and applied as an input for the pulse contour analysis. Before anaesthetic induction, baseline measurements were performed during spontaneous breathing. Mean cardiac output was estimated from the mean of four thermodilution measurements.

After this series of measurements anaesthesia was induced and maintained with sufentanil (initial doses 7·5 μg kg^{-1} and continuous infusion of 3·75 μg kg^{-1} h^{-1} respectively). For muscle relaxation 0·1 mg kg^{-1} pancuronium bromide was given.

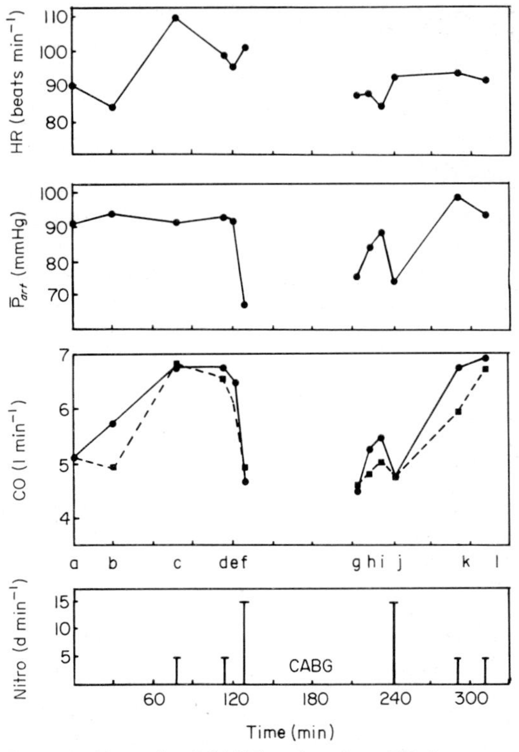

Figure 5 Example of CABG patient data. HR, heart rate; $\bar{P}_{art}$, mean arterial blood pressure; CO, cardiac output for thermodilution (dash line) and pulse contour (solid line); nitro, nitroprusside in drops per min; a . . . l, defined periods as explained in the text.

All patients were ventilated with a Siemens Servo 900B ventilator and a oxygen/air mixture ($FiO_2 = 0{\cdot}5$) at a rate of 10 cycles min^{-1}.

For comparison the thermodilution and pulse contour measurements were performed during haemodynamically stable conditions. This means that no surgical or anaesthetic interventions were carried out during these short periods.

Monitoring of cardiac output

To demonstrate the monitoring capabilities of the system a simplified copy of the screen lay out is given in Fig. 4. The data are from a patient undergoing a CABG operation. In this example the pulse contour method was validated by series of four measurements at different times over the displayed interval. Since no difference was found between the two methods, the trend displayed by the pulse contour method can be considered reliable. The reduction in systolic and diastolic arterial pressure after

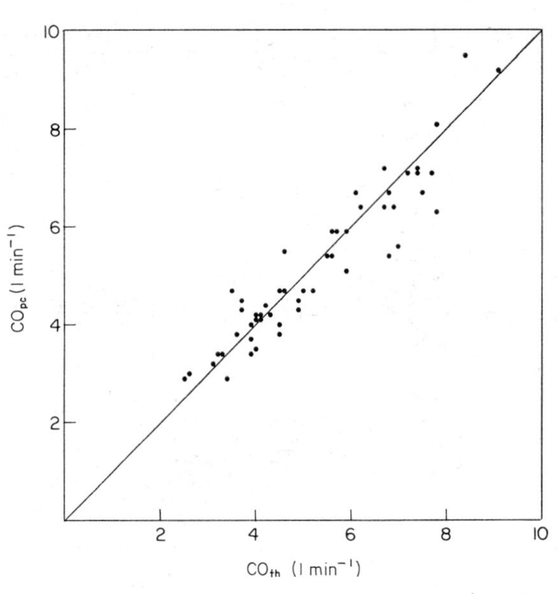

Figure 6 Correlation of cardiac output measured by the thermodilution method (CO_{th}) and pulse contour (CO_{pc}). a=0·31; b=0·94; r=0·94; n=64. Also given is the line of identity.

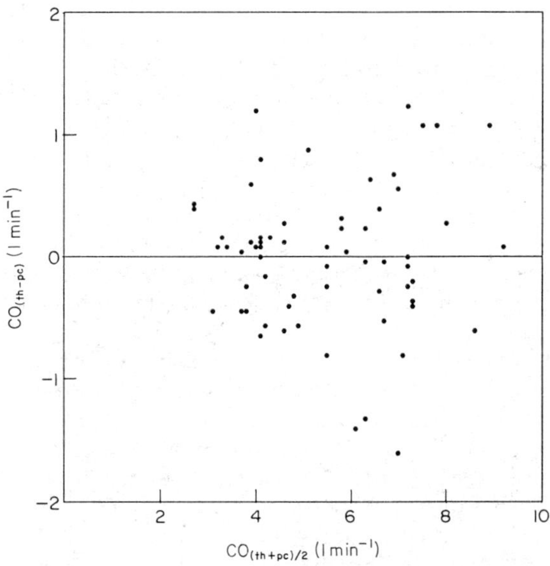

Figure 7 Scattergram for the difference between CO_{th} and CP_{pc} against the mean for both methods. SD=0·54 l min^{-1} or 10·6%, n=64.

the induction of anaesthesia did not coincide with a decrease in cardiac output. Hence, we conclude that, in this patient, induction of anaesthesia caused peripheral vasodilatation, indicated by the total peripheral resistance (TPR display). After anaesthesia had been induced, pressure and cardiac output ran almost parallel.

Validation of the system (pulse contour)

An example of data collected with the system at specific intervals is given in Fig. 5. An initial series of measurements during spontaneous breathing was used to calibrate the system (a). At 11 times during cardiac surgery, a quadruple series of thermodilution measurements was taken, as follows: (b) 3 min after induction of anaesthesia and muscle relaxation; (c) immediately after sternotomy; (d) 15 min before bypass; (e) 10 min before bypass; (f) 5 min before bypass; (g) 3 min after bypass; (h) 8 min after bypass; (i) 13 min after bypass; (j) 18 min after bypass; (k) 3 min after sternal fixation; and (l) end of operation.

During the pre-bypass period, (a) to (e), blood pressure was stable, whereas cardiac output was changing considerably. This indicates substantial changes in total peripheral resistance. Without cardiac output measurements by pulse contour or thermodilution such an episode would have been unnoticed.

A comparison of cardiac output estimated by pulse contour and thermodilution at different moments during cardiac surgery in seven patients is shown in Figs 6 and 7. For each patient the first series of four thermodilution estimates was used to calibrate the pulse contour. This calibration point was removed from the data set. Fig. 6 shows a good overall correlation of $r=0{\cdot}94$ between the methods and a regression slope close to 1. The difference between the methods plotted against the mean of both methods (Fig. 7) was 11%. The data suggest that the difference is related to mean flow; a low flow seems to be estimated more accurately than a high flow. This favours a good estimate during shock and facilitates management of these patients.

If we use a 'worst case' approach for the pulse contour method, assuming thermodilution to be the reference standard and, hence, all scatter of the difference between the methods to be due to pulse contour, 95% of all data are within 22% of the reference values (22% = 2 SD).

Remarks and conclusions

The practical concern is how much reliance can be placed on the pulse contour method for clinically monitoring cardiac output. We, unfortunately, do not have an absolute measure of cardiac output and are left with a comparison with another technique, the choice of which is based on our experience with the commonly used thermodilution technique[1,4,6,23]. However, a comparison between the pulse contour method and thermodilution method is not the ideal basis on which to assess the accuracy of the two techniques. We found a SD for the difference between the two methods of 10·6% (Fig. 6), whereas previous studies yielded a difference of 15%, compared with dye dilution[13], and of 19%, compared with single thermodilution estimates[12]. We expect that averaging of four thermodilution estimates and a more accurate pulse contour analysis both contribute to the smaller scatter of our data.

The difference of 10·6% could not be explained by the repeatability scatter of either method. The SD for repeated measurements of pulse contour during spontaneous breathing was 2·5% and 1·5% during mechanical ventilation, and for the four point averages of the thermodilution the SD was 7% and 3%, respectively[6]. Based on these small SDs we would expect an SD for the difference between the methods of 7·5% and 3·5%, respectively. That we found a higher SD in practice can only be explained by systematic errors in one or both methods. These systematic errors may be dependent on the patient's condition. Our data do not allow us to differentiate between systematic errors in the methods.

We conclude that the corrected pulse contour method, as reported here, followed changes in cardiac output accurately on a beat-to-beat basis even when blood pressure, heart rate and total peripheral resistance changed rapidly and substantially.

References

[1] Jansen JRC, Schreuder JJ, Bogaard JM, v Rooyen W, Versprille A. The thermodilution technique for the measurement of cardiac output during artificial ventilation. J Appl Physiol 1981; 51: 584–91.

[2] Snyder JV, Powner DJ. Effects of mechanical ventilation on the measurement of cardiac output by thermodilution. Crit Care Med 1982; 10: 677–82.

[3] Stevens JH, Raffin TA, Mihm FG, Rosenthal MH, Stetz CW. Thermodilution cardiac output measurement. Effect of the respiratory cycle on its reproducibility. J Am Med Assoc 1985; 253: 2240–2.

[4] Jansen JRC, Versprille A. Improvement of cardiac output estimation by the thermodilution method during mechanical ventilation. Int Care Med 1986; 12: 71–9.

[5] Okamoto K, Komatsu T, Kumar V *et al*. Effects of intermittent positive-pressure ventilation on cardiac output

measurements by thermodilution. Crit Care Med 1986; 14: 977–80.

[6] Jansen JRC, Schreuder JJ, Settels JJ, Kloek JJ, Versprille A. An adequate strategy for the thermodilution technique in patients during mechanical ventilation. Int Care Med 1990; in press.

[7] Warner HR, Swan JC, Conolly D, Tompkins RG, Wood EH. Quantitation of beat-to-beat changes in stroke volume from the aortic pulse contour in man. J Appl Physiol 1953; 5: 495–507.

[8] Alderman EL, Branzi A, Sanders W, Brown BW, Harrison DC. Evaluation of pulse contour method of determining stroke volume in man. Circulation 1972; 46: 546–58.

[9] Starmer CF, McHale FA, Cobb FR, Greenfield JC. Evaluation of several methods for computing stroke volume from central aortic pressure. Circulation 1973; 33: 139–48.

[10] Verdouw PD, Beaune J, Roelandt J, Hugenholtz PG. Stroke volume from central aortic pressure? A critical assessment of the various formulae as their clinical value. Basic Res Cardiol 1975; 70: 377–89.

[11] Tajimi T, Sunagawa K, Yamada A *et al.* Evaluation of pulse contour methods in calculating stroke volume from pulmonary artery pressure curve (comparison with aortic pressure curve). Eur Heart J 1983; 4: 502–11.

[12] Wesseling KH, Purschke R, Smith NT, Nichols WW. Continuous monitoring of cardiac output. Medicamundi 1976; 21/2: 78–90.

[13] Wesseling KH, Wit B de, Weber JAP, Smith NT. A simple device for the continuous measurement of cardiac output. Its model basis and experimental verification. Adv Cardiovasc Phys 1983; 5 (II): 16–52.

[14] Langewouters GJ, Wesseling KH, Goedhard WJA. The static elastic properties of 45 human thoracic and 20 abdominal aortas in vitro and the parameters of a new model. J Biomechanics 1984; 17: 425–35.

[15] Ganz W, Swan HJC. Measurement of bloodflow by thermodilution. Am J Cardiol 1972; 29: 241–6.

[16] Weisel RD, Berger RL, Hechtman HB. Measurement of cardiac output by thermal dilution. New Engl J Med 1975; 292: 67–72.

[17] Levett JM, Replogle RL. Thermodilution cardiac output: a critical analysis and review of literature. J Surg Res 1979; 27: 392–404.

[18] Runciman WB, Ilsley AH, Roberts JG. An evaluation of thermodilution cardiac output measurement using the Swan-Ganz catheter. Aaesth Intens Care 1981; 9: 208–20.

[19] Wessel HU, Paul MH, James GW, Grahn AR. Limitations of thermal dilution curves for cardiac output determinations. J Appl Physiol 1971; 30: 643–52.

[20] Nelson LD, Houtchens BA. Automatic vs manual injections for thermodilution cardiac output determinations. Crit Care Med 1982; 10: 190–2.

[21] American Edwards Laboratories COM-2. Cardiac Output Computer Operations Manual 1989.

[22] Bassingthwaighte JB, Knopp TJ, Anderson DU. Flow estimation by indicator dilution. (Bolus injection): Reduction of errors due to time-averaged sampling during unsteady flow. Circ Res 1970; 27: 277–91.

[23] Jansen JRC, Bogaard JM, Versprille A. Extrapolation of thermodilution curves obtained during a pause in artificial ventilation. J Appl Physiol 1987; 63: 1551–7.

[24] Zierler KL. Circulation times and the theory of indicator-dilution methods for determining blood flow and volume. In: Handbook of physiology. Circulation. Washington, DC: Am Physiol Soc, Vol I, Chap 18, 1962: 585–615.

Radionuclide methods for cardiac output determination

F. M. Fouad-Tarazi and W. J. MacIntyre

The Cleveland Clinic Foundation, Cleveland, Ohio, U.S.A.

KEY WORDS: Radionuclide angiography.

For repeated determinations of cardiac output, radionuclides have provided a safe, reliable and minimally invasive approach. The techniques used are (1) the first-pass dilution method and (2) the derivation of stroke volume from end-diastolic and end-systolic ventricular volumes. Each of the two techniques offers specific advantages, thus, both may be used clinically according to the type of evaluation that is required. Their development has been helped by the availability of adequate radiopharmaceuticals as well as highly efficient precordial detection devices, computer systems, and data handling 'software'. The associated patient exposure to radioactivity has not precluded clinical application because of the small doses used and the protective measures applied.

Introduction

Repeated determinations of systemic haemodynamics are often necessary for effective follow-up of patients. Both short-term and long-term follow-up of patients with cardiovascular disease required sequential evaluation of cardiac output, systemic resistance, blood volume and distribution, as well as indices of cardiac performance. These indices may be altered positively or negatively by the natural history of the disease and/or by therapeutic intervention. To fulfil this haemodynamic follow-up, several non-invasive methods have been developed in two areas: cardiac output determination and blood pressure measurement. Taking into consideration that heart rate could always be recorded non-invasively by ECG, it became practical to assess a large number of haemodynamic indices non-invasively by derivation from cardiac output, heart rate, and blood pressure[1].

For repeated determinations of cardiac output, radionuclides have provided a safe, reliable, and minimally invasive approach. The techniques used are (1) the first-pass dilution method and (2) the derivation of stroke volume from end-diastolic and end-systolic ventricular volumes calculated either on a beat-to-beat basis or from equilibrium radionuclide angiography.

Address for reprints: Fetnat M. Fouad-Tarazi, M.D., The Cleveland Clinic Foundation, Department of Heart and Hypertension Research, One Clinic Center FF10-2b, 9500 Euclid Avenue, Cleveland, OH 44195–5069, U.S.A.

First-pass radionuclide method for determination of cardiac output

This method is based on the analysis of the dilution curve of the tracer 'bolus' as it crosses the heart. The technique utilizes the Hamilton formula[2]. In this, the dilution of the injected substance is determined as a function of time. Since the resultant dilution curve is necessarily dependent upon the volume of flow, cardiac output may be estimated in litres per unit time.

HISTORY

In 1950, Nylin and Celander[3] utilized radiophosphorus-tagged red cells to estimate cardiac output; the dilution was determined by serially sampling arterial blood at short intervals; radioactivity was determined by beta particle assay. In 1951, MacIntyre *et al.* outlined a method for the determination of cardiac output by injection of iodinated (^{131}I) human serum albumin[4]; the method depended on the continuous recording of the dilution curve utilizing gamma ray assay (external gamma detector of high sensitivity).

The use of ^{131}I dominated this field from 1946 to 1963. This was then followed by an era (1963 to present) of ^{99m}technetium (^{99m}Tc) use. In contrast to ^{131}I, which is a man-made radionuclide of a naturally nonradioactive element, technetium had never existed until Emilio Segre and Seaborg produced and identified it in 1937[5].

Progress occurred through the years and led to remarkable improvement in the recording and acquisition systems. In 1962, Folse and Braunwald[6] used a single precordial detector to record radioactivity from a radionuclide indicator injected directly into the left ventricle, and subsequently calculated left ventricular volume. The site of injection of radionuclide was later changed to the vena cava in an attempt to make the procedure less invasive[7,8]. Injections into peripheral veins were adopted later.

The single probe continued to be used by some centres[9,10] for the first pass studies and gated blood pool scanning. However, the development of the scintillation camera by Hal Anger in the 1950s[11] eventually allowed depiction of the spatial distribution of radioactivity within each component of the central circulation over time. Further improvement in this imaging technique paralleled enlargement of the field of view of the camera.

Finally, booming computer development and growth in the field of data processing and display devices have stimulated rapid application of radionuclide-dependent diagnostic tests to clinical cardiology.

RADIOISOTOPE

^{99m}Technetium-labelled human serum albumin (^{99m}Tc-HSA) and ^{99m}Technetium pertechnetate tagged to red blood cells (^{99m}Tc-RBC) have been utilized. ^{99m}Technetium has several characteristics that justify its usefulness in clinical medicine: (a) it decays by isomeric transition which diminishes the radiation dose compared with beta-emitting radionuclide; (b) its half-life is only 6 h, which allows its repeated use within short periods of cardiovascular physiologic evaluation; and (c) its keV photon emission is 140, which allows tissue penetration and therefore external detection without using large amounts of lead shielding of the radiation detectors. The isotope is injected via a peripheral vein; a rapid-flush technique[12,13] provides delivery of the radioactive material to the heart as a bolus. It is essential that the radioisotope remains intravascular during both acquisition of the dilution curve and recording of the equilibrium phase (see 'instruments and acquisition system'). The stability of the bond between ^{99m}Tc and HSA has been tested[13] and shown to be satisfactory both in vitro and in vivo[13,14]. In vitro assessment revealed a binding yield of 96% after 30 min of tagging. In vivo studies revealed 95% binding in plasma samples obtained immediately after the cardiac output study (10 min after injection of the freshly tagged material). Moreover during the 'final dilution' portion of the cardiac output determination (see below), these counts were stable with a slope of 0·86 counts min^{-1} or 4·33 counts. 5 min^{-1}. Red cells labelled in vitro have also been used for other types of first-pass studies[15–18]. We have adapted the in vivo RBC labelling method to the measurement of cardiac output without loss of quality or accuracy of the calculations. More recently, Kelback has reported a modified method to label red blood cells in vitro with ^{99m}Tc while avoiding centrifugation[19]; labelling yields were 95% and in vivo decay showed a high stability with a mean biological half-life of 11 h.

INSTRUMENTS AND ACQUISITION SYSTEM

To record the data necessary for calculation of cardiac output, we have used a portable scintillation camera with a medium-sensitivity, low-energy collimator for precordial recording of the radiotracer's passage across the chambers of the heart. The camera head is positioned in a left anterior oblique position at 30–45°, parallel to the longitudinal body axis and tilted 0–5° upward to help visualize the subclavian veins.

The camera output is transferred to a storage system, with individual frames stored on magnetic tape. Recording, storing and playback functions are effected by an offline computer. This procedure permits recording and storage of sequential full frames of the multichannel analyser at 0·5 s intervals without loss of transfer time from the gamma camera. In heart failure, 1 s intervals are necessary to visualize the full dilution curve.

Cardiac output by radionuclide technique has been also determined by the collimated scintillation probe[21,22]. The scintillation probe provides a convenient portable means of evaluating cardiac output at the patient's bedside; however, its use is limited by the difficulty of appropriate positioning of the probe relative to the cardiac chambers.

Procedure

With ^{99m}Tc, the usual adult dose is 10–20 mCi; for children 0·1 mCi kg^{-1} has been suggested[20]. In our practice, a bolus of 8–12 mCi ^{99m}Tc-HSA is used in the case of a single output determination. In the event of two output determinations at the same sitting, a dose of 4 mCi is used for the first study and 8 mCi for the second. A background frame of 60-s duration is obtained prior to the second injection. The counts during this second study are then corrected for background. Whole body radiation

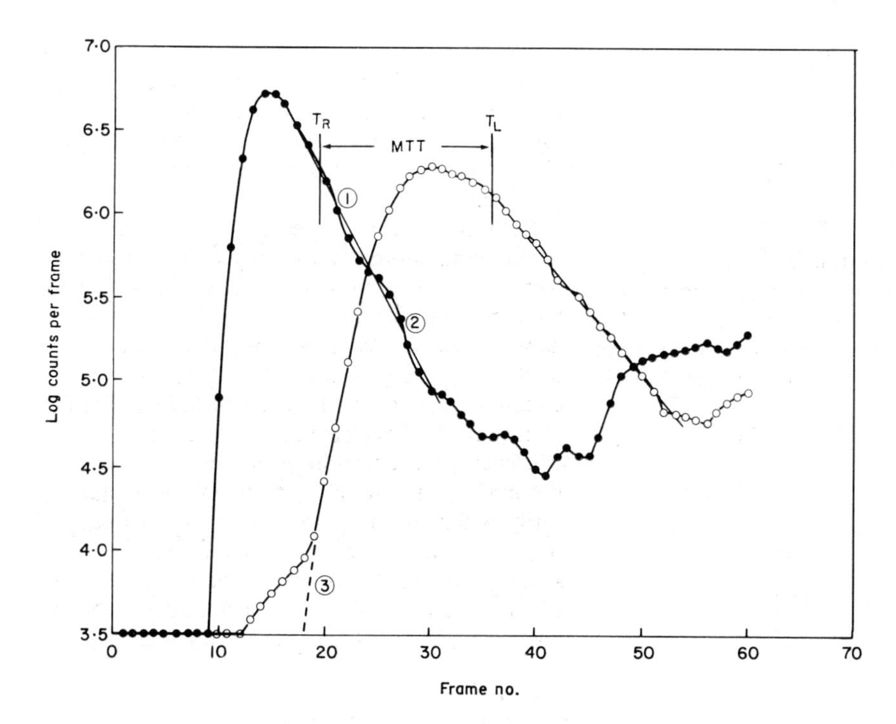

Figure 1 Dilution curves from right ventricular (●) and left ventricular (○) regions of interest. ① ② and ③ represent extrapolation lines. MTT indicates mean transit time measured by centroid method (T_R and T_L, adapted from [52]).

dose originating from 1 mCi was calculated to be <0·018 rad.

It is highly desirable that the radionuclide enter the heart as a compact bolus. Failure to deliver a discrete bolus rapidly may lead to erroneous determination of the descending limb of the dilution curve; this error becomes increasingly severe in each downstream chamber. The ideal injection technique[23] is via a catheter in the superior vena cava at its entrance into the right atrium, time of injection corresponding to the beginning of systole (to assure best mixing) and at the beginning of inspiration (to facilitate cardiopulmonary transit). This technique is, however, often impractical and most investigators use a percutaneous venous injection preferably into the right basilic vein at the antecubital fossa[13].

The bolus is flushed with 20 ml normal saline over 3 s. During the following 30 s, 60 frames are collected for 0·5 s periods on the Datastore and stored sequentially on the magnetic tape. One-second intervals are used in heart failure patients as explained above. An additional 20 frames are then collected during equilibrium, each for 30 s (total of 10 min), and are stored on the tape for calculation of final dilution.

DATA ANALYSIS AND CALCULATION OF CARDIAC OUTPUT

Measurements of cardiac output by the indicator dilution technique (radioactive of non-radioactive) depend on the Stewart-Hamilton principle[2]. The cardiac output is calculated by observing the time-activity (external monitoring by the scintillation camera) profile of the tracer after its injection into the circulatory system (Fig. 1). The data are acquired by flagging the region of interest (ROI) in the left ventricle on video playback[21,24]. The cardiac output is calculated from the formula:

$$R = \frac{I}{\int_o^\infty c_t dt}$$

where F = blood flow, I = dose of tracer, c = blood concentration of tracer during the bolus passage, t = time.

$c_t \cdot dt$ represents the area under the dilution curve; indeed, if we compare 'in heart' dilution with 'external' (transthoracic) dilution

$$\int_o^{\infty} \frac{c_{(t)}dt}{Ceq} = \int_o^{\infty} \frac{q_{(t)}dt}{Qeq}$$

where c = blood concentration of tracer during the bolus passage, q = mean height of the tracer curve during the bolus passage as recorded externally, Ceq = blood concentration of tracer after complete mixing, Qeq = height of the tracer curve after complete mixing.

Thus,

$$F = \frac{I}{\int_o^{\infty} q_{(t)}dt \; X \; \frac{Ceq}{Qeq}}$$

$$F = \frac{Qeq}{\int_o^{\infty} q_{(t)}dt} \; X \; \frac{I}{Ceq}$$

It should be noted that I/Ceq represents the distribution volume (Vd) of the tracer. Since $\int_o^{\infty} q_{(t)}dt$ represents the area (A) under the externally recorded passage of the tracer bolus, and I/Ceq equals the volume of dilution, Vd, the equation for calculation of flow becomes:

$$F = \frac{Qeq}{A} \, X \, V_d$$

Measurement of the area under the curve is problematic because of the recirculation peak superimposed on the tail of the dilution curve. To correct for recirculation, a semilogarithmic extrapolation of the tail of the first curve[25–27] has been used; alternatively, a gamma variate function with least square analysis of the up-slope of the curve and a portion of the downslope of the curve could be applied[28–30].

The numerator I represents the total quantity of isotope injected. It is equal to the equilibrium concentration of isotope in the ROI (the area of LV monitored by the external probe) multiplied by the blood volume. The equilibrium concentration of isotope Qeq can be obtained from the height of the equilibrium activity curve (30 s static image) obtained at the end of 10-min recording of tracer mixing in the circulation[13,19,20,31]. Since the same fraction of the total blood volume is involved in the assessment of both Ceq and A, this fraction of the blood volume (volume of ROI) does not appear in the equation.

Kelback *et al.*[19] stressed the importance of background subtraction during processing of the stored data; they chose for background an area in the region of least activity just inferolateral to the apex of the heart. Whereas this procedure may add to the refinement of the method, the statistics of counting depend on the concentration of radiopharmaceutical reaching the heart as well as on the efficiency of the detector system. In most instances, the maximum counting rate is achieved satisfactorily within the field of view with the usage of the recommended ^{99m}Tc dosages. An excellent signal/'noise' ratio (cardiac ROI counts/background counts) is usual since background levels are low during first pass radionuclide angiography.

VALIDATION OF THE METHOD

Validation of the first pass radionuclide method for cardiac output determination has been reported on several occasions. In 1979, we reported that the correlation between radionuclide cardiac output and simultaneous dye dilution cardiac output was acceptable whenever the radionuclide material reached the heart as a discrete bolus and there was no circulatory delay in the veins at the thoracic inlet[13]. A correlation of 0·98 ($P<0{\cdot}001$) was reported by Kelback *et al.*[19] between first-pass radionuclide cardiac output and conventional tracer dilution on arterial sampling of 14 patients with heart disease of various origins, indicating that the radionuclide cardiographic technique is a valuable tool for non-invasive evaluation of patients.

Cardiac output derived from stroke volume calculated from ECG-gated radionuclide ventriculography

Equilibrium radionuclide ventriculography was used to determine LV end-diastolic and end-systolic volumes as early as 1978[32,33]. The method for LV volume determination was examined further by other investigators[34]; correction factors were introduced and cardiac output was calculated as:

$$(EDV–ESV) \; X \; HR = SV \; X \; HR = CO$$

The method was also applied to children[34], and measurements were validated against cineangiographically determined LV volumes and cardiac outputs. Radionuclide LV volume determination has advantages over cineangiographically determined images. The count-based methods are less sensitive to ventricular contour drawing[35]; they are said to be free of assumptions about the actual shape of the ventricle being imaged[34,36–38]. However, there are disadvantages inherent in the radionuclide methods. On the one hand, some of these methods rely on assumptions about the average attenuation of gamma rays emitted from the heart chambers[39,40], others require blood sampling[39–41]. Newer techniques require the placement of a source of radioactivity in the patient's oesophagus for correction of attenuation, an uncomfortable procedure for the patient[42].

These correction factors were criticized for several potential pitfalls[43]; the photon pathway from the centre of the left ventricle to the scintillation camera crystal involves passage through a number of different tissues with varying densities including blood muscle, lung, containing air, and bones. Therefore, the linear attenuation coefficient of water which Links et al.[39] used for the attenuation corrections is not appropriate. It has been reported that scattered photons can cause a 50% overestimation of the actual activity[44]. Nickoloff *et al.*[43] developed a computer program to convert automatically CT numbers obtained in routine patient scanning into linear attenuation coefficients at ^{99m}Tc energy. Such data indicated that an average value for the effective linear attenuation coefficient is 0·12 g cm^{-1}. Moreover, the study of Nickoloff *et al.*[43] indicated that both the effective linear attenuation coefficient and attenuation path length are patient-dependent because of varying amounts of air in the lung and varying amounts of body tissue. Therefore, the use of a single value for all patients would result in left ventricular volume errors that ranged from −27% to +35% in the study of Nickoloff *et al.*[43] This study has also shown that although the oesophageal source approach[42] takes into consideration the patient-to-patient variability, it is position-dependent. It tends to overestimate the attenuation because the source is behind the heart and the average attenuation path length is 66% further from the chest wall than the centre of the left ventricle.

In 1984, Nichols *et al.*[45] introduced a new method, free from assumptions of ventricular shape and correction for radiation attenuation, that needs no additional time or expertise. The stroke volume obtained by this method correlated with thermodilution-determined stroke volume ($r = 0{\cdot}98$).

Overall, the accuracy of these volumetric radionuclide methods has been influenced by all factors that affect the accuracy of ejection fraction measurements, including background subtraction[46–48], irregularities of heart beat[49], respiration[50], as well as the algorithm for cardiac phantom[51].

Discussion

The use of imaging techniques for quantitative measures of cardiac output allowed the non-invasive quantitative evaluation of cardiovascular dynamics. These evaluations take a short time and analysis of data is fast because of availability of adequate computing 'hardware' and analysis algorithms.

The first pass dilution technique has several advantages over the volumetric method (Table 1). In addition to measuring cardiac output, it allows calculations of preliminary mean transit time and derivation of cardiopulmonary blood volume[52]. The ratio of cardiopulmonary volume to concurrently measured total blood volume provides an indirect indication of venous tone. Moreover, comparison of cardiac output determined from the right ventricular region of interest with the cardiac output determined from the left ventricular region of interest provides an internal check for accuracy. Also, calculations are not influenced by attenuation factors since the same region of interest is used for calculation of both the 'area under the dilution curve' and the 'final dilution of the radiopharmaceutical'. Despite its simplicity, the method has encountered delay in its widespread use, mainly for technical reasons: on the one hand, determination of the precise dose of radiopharmaceutical reaching the heart for mixing was difficult to calibrate until external calibration methods developed; on the other hand, first pass radionuclide angiography has other limitations compared with the radionuclide volumetric method for measuring cardiac output (Table 2). First, camera positioning is more critical during the former because of lack of radioactivity in the cardiopulmonary area prior to injection. Since the latter method involves equilibrium circulation of the radioisotope, camera positioning is less critical and separation of right and left ventricle and of atria and ventricles may be controlled more precisely. A second advantage of the radionuclide

Table 1 Advantages of first pass cardiac output vs ventriculography cardiac output

First-Pass	Ventriculography
Non-geometrical dilution	Non-geometrical but depends on LV boundaries
Not influenced by attenuation factors	Influenced by attenuation factors
Not influenced by arrhythmias	Influenced by arrhythmias
May be applied to normal and abnormal hearts	More errors in abnormal hearts due to low ejection fraction
Allows measurement of MTT and CPV	Not applicable
May use internal check for accuracy of measurement (RV-CO vs LV-CO)	Not applicable; RV ejection fraction is more open to errors
Faster computer analysis	Slower due to additional measurements to locate the heart centre

Table 2 Advantages of ventriculography cardiac output vs first-pass cardiac output

First-pass	Ventriculography
Requires blood volume measurement	Does not require blood volume measurement
Camera positioning is critical	Camera positioning is not critical
Radiotracer bolus injection should be compact and discrete	Radiotracer in the heart is imaged at equilibrium
Number of cardiac output measurements is limited during a single study	Number of cardiac output measurement during a single study is more generous
Not applicable	Also, allows derivation of ejection fraction, ventricular wall motion and peak rates of LV ejection and filling

volumetric method is the unlimited number of measurements that can be achieved provided the ^{99m}Tc tagging remains stable; thus, several manoeuvers or interventions may be assessed during the same study in the same individual. In contrast, the number of cardiac output measurements during first pass radionuclide angiography is restricted to three because the amount of radioactivity that an individual may be exposed to is limited. Another advantage of the volumetric method is the possibility of deriving other indices from the left ventricular volume (counts) change after radiotracer has equilibrated in the vascular stream. The gated blood pool scintiscans allow measurement of left ventricular ejection fraction[53], left ventricular peak ejection rate, and left ventricular peak filling rate[54]; moreover, they permit determination of myocardial wall motion.

The equipment used for imaging is critical for both the first pass and the equilibrium radionuclide angiography methods. The scintillation probe provides a convenient portable means of evaluating cardiac output at the patient's bedside; however, the position of probe placement relative to the left ventricle is critical. The wide field of view of the scintillation camera allows viewing of the whole heart and diminishes the frequency of misplacement of the probe relative to the cardiac location. Moreover, the wide field of view allows visualization of the upper thoracic veins (subclavian and superior vena cava); this monitoring of the passage of the radioactive bolus via these veins proved to be of major importance in our experience, during acquisition and analysis of the dilution curves[13]. 'Venous delay' at the level of the subclavian veins influenced the accuracy of cardiac output determination by this method irrespective to its cause. Oldendorf *et al.*[12], Ashburn[14] and Lane *et al.*[55] described methods of rapid bolus delivery with intravenous injection, but anatomical variations may still produce delays of the radioactivity bolus flow and this may only be detected by direct imaging during the first pass study. Finally, the separation between the left ventricular (LV) image and the right ventricular image is easier to achieve by using the large field of view camera, a prerequisite for purer dilution curves, necessary for some calculations derived from the right and left dilution curves, e.g., the pulmonary mean transit time and cardiopulmonary blood volume. This discussion does not exclude the use of the single probe for cardiac output calculations; indeed, it has been our experience that cardiac output calculated from the 'total heart region' visualized by the wide-angle camera[54] correlated significantly with the cardiac output obtained by either the dye-dilution method or from the LV region of interest defined by radionuclide (^{99m}Tc) technique. This good correlation suggested the possibility of using a smaller dose of radiotracer for a single study and is of clinical value in situations where other measurements (such as MTT and CPV) are not sought.

In conclusion, radionuclides provide a safe, reliable and minimally invasive approach for cardiac output determination. Both first-pass dilution and volumetric methods are useful clinically, each having specific applications according to the type of evaluation required. Further refinement of the present techniques will increase their usefulness for patient care and evaluation of cardiovascular physiology.

References

[1] Tarazi RC, Ibrahim MM, Dustan HP, Ferrario CM. Cardiac factors in hypertension. Circ Res 1974; 34 (Suppl I): 213–21.

[2] Hamilton WF, Riley RL, Attyah AM *et al.* Comparison of the fick and dye injection methods of measuring the cardiac output in man. Am J Physiol 1948; 153: 309–32.

[3] Nylin G, Celander H. Determination of blood volume in the heart and lungs and the cardiac output through the injection of radiophosphorus. Circulation 1950; 1: 76–83.

[4] MacIntyre WJ, Pritchard WH, Eckstein RW, Friedell HL. The determination of cardiac output by a continuous recording system utilizing iodinated (^{131}I) human serum albumin I. animal studies. Circulation 1951; 4: 552–6.

[5] Wagner HN Jr. The development of cardiovascular nuclear medicine. In: Strauss HW, Pitt B, James AE Jr., eds. Cardiovascular nuclear medicine. St Louis: The C. V. Mosby Co, 1874.

[6] Anger HO. Scintillation camera. Rev Sci Instr 1958; 29: 27–33.

[7] Folse R, Braunwald E. Determination of fraction of left ventricular volume ejected per beat and of ventricular end-diastolic and residual volumes. Circulation 1962; 25: 674–85.

[8] Donato L. Selective quantitative radiocardiography. Prog Cardiovasc Dis 1962; 5: 1.

[9] Donato L. Basic concepts of radiocardiography. Semin Nucl Med 1973; 3: 111–30.

[10] Man in 't Velt AJ, Wenting GJ, Verhoeven RP, Schalekamp MADH. Quantitative radiocardiography by single-probe counting using ^{99m}technetium albumin. Neth J Med 1978; 21: 166–75.

[11] van Brummelen P, Man in 't Veld AJ, Schalekamp MADH. Hemodynamic changes during long-term thiazide treatment of essential hypertension in responders and nonresponders. Clin Pharmacol Ther 1980; 27: 328–336.

[12] Oldendorf WH, Kitano M, Shimizu S. Evaluation of a simple technique for abrupt intravenous injection of a radioisotope. J Nucl Med 1965; 6: 205–9.

[13] Fouad FM, Tarazi RC, MacIntyre WJ, Durant D. Venous delay, a major source of error in isotopic cardiac output determination. Am Heart J 1979; 97: 477–84.

[14] Ashburn WL. Videotape applications in cerebral blood flow studies. In: Gilson AJ, Smoak WM, Eds. Central nervous system investigation with radionuclides. Springfield: Charles C. Thomas, 1971: 279.

[15] Steele P, Kirch D, Lefree M, Battock D. Measurement of right and left ventricular ejection fraction in coronary artery disease. Chest 1976; 70: 51–6.

[16] Dymond DS, Grenier RP, Carpenter J, Schmidt DH. First-pass radionuclide angiography via pulmonary arterial catheters. A critical analysis of background components. Radiology 1984; 150: 819–23.

[17] Hurwitz RA, Treves S, Kuruc A. Right ventricular and left ventricular ejection fraction in pediatric patients with normal hearts: first-pass radionuclide angiocardiography. Am Heart J 1984; 107: 726–32.

[18] Harpen MD, Robinson AE. Radionuclide determination of right- and left-ventricular mixing. Eur J Nucl Med 1985; 10: 5–9.

[19] Kelbaek H, Hartling OJ, Skagen K, Munck O, Henriksen O, Godtfredsen J. First-pass radionuclide determination of cardiac output: An improved gamma camera method. J Nucl Med 1987; 28: 1330–4.

[20] Quantitative nuclear cardiography. Pierson RN Jr, Kriss JP, Jones RH, MacIntyre WJ, Eds. New York: John Wiley & Sons Inc, 1975.

[21] VanDyke D, Anger HO, Sullivan RW. Cardiac evaluation from radioisotope dynamics. J Nucl Med 1972; 13: 585–92.

[22] MacIntyre WJ, Storaasli JP, Krieger H, *et al.* ^{131}I-labeled serum albumin: Its use in the study of cardiac output and peripheral vascular flow. Radiology 1952; 59: 849.

[23] Freedman GS. Radionuclide angiocardiography in the adult. In: Strauss HW, Pitt B, James AE Jr, Eds. Cardiovascular nuclear medicine, St. Louis: The C. V. Mosby Co, 1874.

[24] Bitter F, Besch W, Schafer N *et al.* Integrierte Herz-Kreislaufanalyse mit Hilfe der quantitativen Funktionsszintigraphie. In: Horst W, Ed. Frontiers of nuclear medicine, Berlin: Springer-Verlag, 1971: 250–61.

[25] Hamilton WF, Moore JW, Kinsman JM, Sparling RG. Studies on the circulation; IV. Further analysis of the injection method, and of changes in hemodynamics under physiological and pathological conditions. Am J Physiol 1931, 32; 99: 534.
[26] Hamilton WF. Circulation. In: Handbook of physiology. Washington, DC: American Physiological Society 1962: 567–84.
[27] Razzak MA, Botti RE, MacIntyre WJ, Pritchard WH. Consecutive determination of cardiac output and renal blood flow by external monitoring of radioactive isotopes. J Nucl Med 1970; 11: 190–5.
[28] Thompson HK, Starmer CF, Whalen RE, McIntosh HD. Indicator transit time considered as a gamma variate. Circ Res 1964; 14: 502–15.
[29] Starmer CF, Clark DO. Computer computations of cardiac output using the gamma function. J Appl Physiol 1970; 28: 219–20.
[30] Steadham RE, Blackwell LH. A new method for the determination of the area under a cardiac output curve. IEEE Trans Biomed Eng 1970; 17: 335.
[31] Holman BL, McNeil BJ, Adelstein SJ. Quantitative tracer kinetics. In: Gottschalk A, Potchen JE, Eds. Diagnostic nuclear medicine. Baltimore: Williams & Wilkins Co, 1976: 116–25.
[32] Bingham J, Taroli E, Alpert N *et al.* Determination of left ventricular end-diastolic volume from gated cardiac images: comparison with contrast ventriculography. J Nucl Med 1978; 20: 667–8.
[33] Dehmer G, Firth B, Lewis S, Willerson J, Hillis L. Direct measurement of cardiac output by gated equilibrium blood pool scintigraphy: validation of scintigraphic volume measurements by a nongeometric technique. Am J Cardiol 1981; 47: 1061–7.
[34] Parrish MD, Graham TP Jr, Born ML, Jones JP, Boucek RJ Jr, Partain CL. Radionuclide ventriculography for assessment of absolute right and left ventricular volumes in children. Circulation 1982; 66: 811–9.
[35] Massie BM, Kramer BL, Gertz EW *et al.* Radionuclide measurement of left ventricular volume: comparison of geometric and count-based methods. Circulation 1982; 65: 725–30.
[36] Slutsky R, Karliner J, Ricci D *et al.* Left ventricular volume by gated equilibrium radionuclide angiography: a new method. Circulation 1979; 60: 556–64.
[37] Dehmer GJ, Lewis SE, Hillis LD *et al.* Nongeometric determination of left ventricular volumes from equilibrium blood pool scans. Am J Cardiol 1980; 45: 293–300.
[38] Clements IP, Brown ML, Smith HC. Radionuclide measurement of left ventricular volume. Mayo Clinic Proc 1981; 56: 733.
[39] Links JM, Becker LC, Shindledecker JG *et al.* Measurement of absolute left ventricular volume from gated blood pool studies. Circulation 1982; 65: 82–91.
[40] Thomsen JH, Patel AK, Rower BR *et al.* Estimation of absolute left ventricular volume from gated radionuclide ventriculograms. A method using phase image assisted automated edge detection and two-dimensional echocardiography. Chest 1983; 84: 6–13.
[41] Harpen MD, Dubuisson RL, Head GB, Parmley LF, Jones TB, Robinson AE. Determination of left-ventricular volume from first pass kinetics of labeled red cells. J Nucl Med 1983; 24: 98–103.
[42] Maurer AH, Siegel JA, Denenberg BS *et al.* Absolute left ventricular volume from gated blood pool imaging with use of esophageal transmission measurement. Am J Cardiol 1983; 51: 853–8.
[43] Nickoloff EL, Perman WH, Esser PD, Bashist B, Alderson PO. Left ventricular volume: physical basis for attenuation corrections in radionuclide determinations. Radiology 1984; 152: 511–5.
[44] Siegel JA, Wu RK. The elusive build-up factor. Med Phys 1982; 9: 614.
[45] Nichols K, Adatepe MH, Isaacs GH *et al.* A new scintigraphic method for determining left ventricular volumes. Circulation 1984; 70: 672–80.
[46] Goris ML, Daspit SG, McLaughlin P, Kriss J. Interpolative background subtraction. J Nuc Med 1976; 17: 744–7.
[47] Narahara KA, Hamilton AW, Williams DL, Gould KL. Myocardial imaging with thallium-201: an experimental model for analysis of true myocardial and background image components. J Nucl Med 1977; 18: 781–6.
[48] Goris JL. Nontarget activities: can we correct for them? J Nucl Med 1979; 20; 1312–4.
[49] Slutsky R. On the analysis of left ventricular volume from gated radionuclide ventriculograms. Chest 1983; 84: 2–3.
[50] Kim BH, Ishida Y, Tsuneoka Y *et al.* Effects of spontaneous respiration on right and left ventricular function: evaluation by respiratory and ECG gated radionuclide ventriculography. J Nucl Med 1987; 28: 173–7.
[51] Cradduck TD, Busemann-Sokole E. Dependence of ejection fraction results on choice of algorithms for a cardiac phantom. Nucl Med Comm 1986; 7: 33–44.
[52] Fouad FM, MacIntyre WJ, Tarazi RC. Noninvasive measurement of cardiopulmonary blood volume. Evaluation of the centroid method. J Nucl Med 1981; 22: 205–11.
[53] Lane SD, Patton DD, Staab EV, Baglan RJ. Simple technique for rapid bolus injection. J Nucl Med 1972; 13: 118–9.
[54] Fouad FM, Houser T, MacIntyre WJ, Cook SA, Tarazi RC. Automated computer program for radionuclide cardiac output determination. J Nucl Med 1979; 20: 1301–7.

First passage radionuclide cardiography for determination of cardiac output: A critical analysis

G. J. WENTING, R. M. L. BROUWER, A. J. MAN IN 'T VELD AND M. A. D. H. SCHALEKAMP

Department of Internal Medicine I, University Hospital Rotterdam 'Dijkzigt', Rotterdam, The Netherlands

KEY WORDS: Radionuclide cardiography, cardiac output, scintillation probe, gamma camera, dye dilution, accuracy, precision.

First-pass radiocardiography, by single probe or gamma camera, has the theoretic potential to be an attractive non-invasive indicator dilution method of measuring cardiac output. Registrations, once programmed, require little time to perform and entail hardly any risk for the subject. At the same time, they are to varying degrees inaccurate. As long as technology is not standardized, each institution that wishes to employ these measurements has to do its own critical validation before results can be accepted. As with any other indicator dilution technique, precision of radiocardiography is served by repetitive measurements. However, radiation dose and disturbing background radioactivity preclude taking multiple measurements within a short period. This holds particularly for the gamma camera. The speed and simplicity of the probe system make this device very suitable for serial evaluation of cardiac function at the bedside. Depending on collimation of the probe and extracardiac background activity, a correction factor has to be derived empirically to avoid overestimation of cardiac output. The major advantage of the gamma camera linked to a data system is that an infinite number of first-pass curves can be obtained from different parts of the heart. Provided that appropriate regions of interest are selected, first-pass studies can yield reasonably accurate and reproducible determinations of cardiac output. In addition, functional image analysis during the equilibrium phase enables calculation of other cardiac variables such as ejection fraction and chamber size. Nevertheless, standardization of 'hardware' and 'software' is imperative.

Introduction

Of the various methods available for the measurement of cardiac output, first passage radiocardiography with a gamma-emitting isotope and external counting has multiple advantages; in fact it has almost a 'fairy-tale' appeal. It makes use of the sound principle of indicator-dilution formulated by Hamilton-Stewart[1]. This analysis of the arterial time-concentration curve following a bolus of an indicator substance has been demonstrated to be accurate, both mathematically and experimentally[2]. However, using external counting of arterial radioactivity, there is no requirement for catheterization of the heart or great vessels. The patient is inconvenienced only to the extent of ongoing venepuncture, thus the existing physiologic state presumably is disturbed minimally as a result of the measurement. In the face of these compelling arguments, one may wonder why external counting methods have had relatively limited application.

Correspondence to: Dr G. J. Wenting, Department of Internal Medicine I, room D 410, University Hospital Rotterdam 'Dijkzigt', Dr Molewaterplein 40, 3015 GD Rotterdam, The Netherlands.

Early first-pass measurements were performed with collimated scintillation detectors placed over the chest[3–5]. Agreement on the accuracy of these measurements using ^{24}Na or ^{131}I albumin as indicators was not unanimous[6–8], and widespread use of the method was hampered by the lack of short-lived isotopes. Nowadays, radioisotopes such as ^{99m}technetium and ^{113m}indium have solved this problem. Parallel to the development of laboratory generators of these isotopes, the gamma-camera was developed, and this became the universal detection system in nuclear medicine, consigning the probe to oblivion.

During the last decade, numerous reports have appeared describing methods for determining cardiac output using radionuclides and the gamma camera[9]. While several studies[10–13] have emphasized the use of gated blood-pool imaging (measurement of cardiac output through determination of ventricular volume which, with heart rate and ejection fraction, yields cardiac output), the first-pass method based on indicator-dilution has been evaluated with more favourable results[14–16]. Nevertheless, discussion and debate continue on the accuracy and reproducibility of the last method[17,18]. It is the purpose of this paper to delineate some of the problems of first-pass radiocardiography.

Quantitative radiocardiography by the single probe

THEORETICAL BACKGROUND

Early first-pass radionuclide cardiac output measurements were performed with collimated gamma-sensitive scintillation detectors placed over the heart. While almost all investigators agreed that proper placement and collimation of the probes were critical to some extent to the success of the method[8], there was no general agreement in practice as to where or how the probes should be placed and how narrow or wide their scanning field should be focused. Given the rapid development of the gamma camera systems, further exploration of the potential value of the scintillation probes was not pursued.

However the scintillation probe is an efficient radiation detector providing high count-rates for good statistical reliability. Their efficiency with regard to photon detection is two orders of magnitude greater than current gamma camera systems. This important feature of the probe enables the use of low single radiation doses (75–150 μCi of ^{99m}technetium) which allow multiple studies to be performed in a short period. Since gamma cameras lack the high count-rate efficiency inherent to probe systems, they can only be used with relatively high radiation doses of radiopharmaceuticals (15–20 mCi). However, in contrast to probes, the gamma camera scans the whole heart and an infinite number of first-pass curves can be obtained from different parts of the heart by use of appropriate software and a 'dedicated' computer.

In principle, however, the cardiac area 'seen' by the detecting gamma-sensitive device (probe or camera) should not affect the measurement of cardiac output[8,19,20]. In this regard, it is important to recollect the theoretical background of the first-pass method. When a quantity (I) of an intravascular indicator is injected rapidly into the right atrium, it will appear downstream in the pulmonary artery and left side of the heart at a concentration (C) that varies with flow (F) and time (t). The time–concentration curve at a downstream sampling point is related by conservation of mass to a constant flow, according to the equation:

$$I = F \int_0^\infty C(t)dt \qquad [1]$$

This equation can be expressed in the form:

$$F = \frac{\text{dilution volume} \times E}{A} \qquad [2]$$

In the case of radiocardiography, monitoring of the time–concentration curve (A) is performed with an external detector. E is the count rate of the detector at the time when dilution volume of the intravascular tracer is measured (blood sample, calibration phase). It should be stressed that equation 2 is valid even when multiple sites of flow are observed.

The important implication, however, with external counting (done by probe or camera) is that all intravascular radioactivity contributing to the final counting rate during calibration should reside within the same circulatory segments that contribute radioactivity during the earlier inscription of the first-pass dilution curve. These theoretical aspects relate to the problem of methodological accuracy, which is discussed below.

As far as the area observed by the probe is concerned, the segments initially contributing activity are presumed to include one or both cardiac chambers, the pulmonary artery, the aorta, or any combination of these, depending upon placement of the detector. The precordial count made at the subsequent equilibrium time clearly also includes activity in large intrathoracic venous segments, the vessels of the chest wall, and the vessels of the myocardium. To the extent that these additional segments contribute, the calibration is invalid, the counting rate is spuriously high, and blood flow is overestimated[18].

The crucial question of whether technical performance of the probe method can be standardized to keep the undesirable background radioactivity negligible, or at a predictable level, has received insufficient attention. In this regard the gamma camera linked to a data system has certain advantages. Regions of interest over the left ventricle, aorta or whole heart can be obtained easily by a dedicated computer. In this way, the influence of radioactivity from extracardial sources on the equilibrium count rate can be reduced, but not excluded[21].

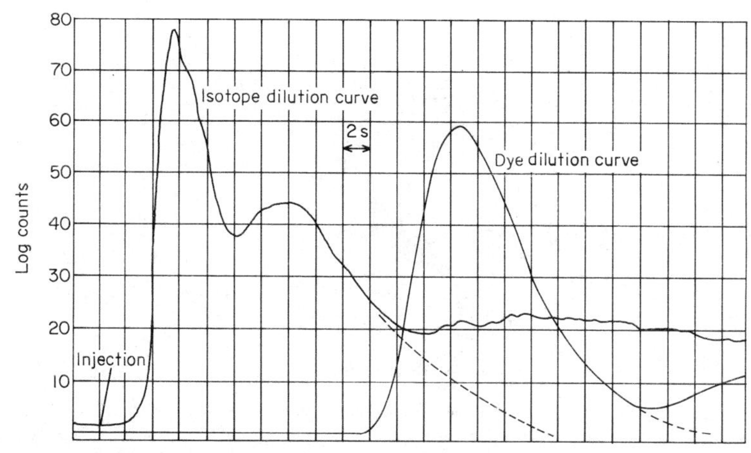

Figure 1 Simultaneous measurements of cardiac output by isotope dilution (scintillation probe) and dye dilution.

A major disadvantage of the gamma camera system, however, is relatively poor curve resolution. As a consequence, doses of radioactivity must be much higher than those needed with the probe. Hence, replicate measurements cannot be obtained without exceeding acceptable radiation exposure. For these reasons we re-evaluated the single probe system using ^{99m}Tc-albumin as a tracer for the serial measurement of cardiac output. A comparison with the classical indocyanine green dilution method was performed in a relatively large population of patients.

COMPARISON WITH DYE DILUTION

Simultaneous cardiac output determinations by precordial single probe counting (CO_{Tc}, ^{99m}technetium albumin as tracer) and by dye (CO_{dye}, indocyanine green) were compared in 38 patients (55 paired measurements). Patients, procedures, equipment and calculations have been described in detail previously[22]. A simultaneous recording of an isotope and dye dilution curve is shown in Fig. 1. Note the two peaks of the isotope dilution curve, corresponding to the passage of the indicator through the right and left heart, respectively. Depending on the localization of the detector over the chest, variable contributions of the 'right' and 'left' heart curve to the total area under the curve were obtained. In order to eliminate the contribution of recirculation of the tracer, the down-slope of the left heart curve was extrapolated. Dye was sampled from the brachial artery, which explains why inscription of the curve occurred later in time. A scatterplot of the results of the two methods is shown in Fig. 2, panel a.

For a description of the relationship between the two methods, linear regression analysis and computation of the product-moment correlation coefficient (r) are often used as an indicator of agreement. Doing so, we found a high correlation coefficient, $r = 0{\cdot}94$, $P < 0{\cdot}001$, between the two methods. Obviously, both methods are linearly related. However, as discussed recently by Bland and Altman[23], this high correlation does not mean that the two methods agree. Looking at the scatterplot and plotting the difference between the two methods against their mean (Fig. 2, panel b), it is clear that the probe method (CO_{Tc}) overestimates cardiac output as measured by dye-dilution (CO_{dye}) over the whole range of values. We found a mean difference (CO_{Tc} minus CO_{dye}) of $+2{\cdot}0\ l\ min^{-1}$ with a 95% confidence interval of the difference, of 0–4 $l\ min^{-1}$.

Such a systematic difference between the two methods has been described before by several investigators[4,6,7] and can be explained by the influence of background radioactivity on the final counting rate during calibration (see above). If this is a consistent bias we can adjust for it with an empirical correction factor. For our probe system and series of measurements the regression equation relating CO_{Tc} to CO_{dye} is: $CO_{dye} = -0{\cdot}02 + 0{\cdot}75 . CO_{Tc}$; in practice this means: $CO_{dye} = 0{\cdot}75 . CO_{Tc}$. Investigators who used a narrow-angle collimator found correction factors close to one[5,24–26], while the collimator used by Donato *et al.*[4] (who found a correction factor of 0·83) had a relatively large aperture. In our study we used a collimator with a rather large aperature, which partly explains why we found a correction factor in the order of 0·75. It should be noted,

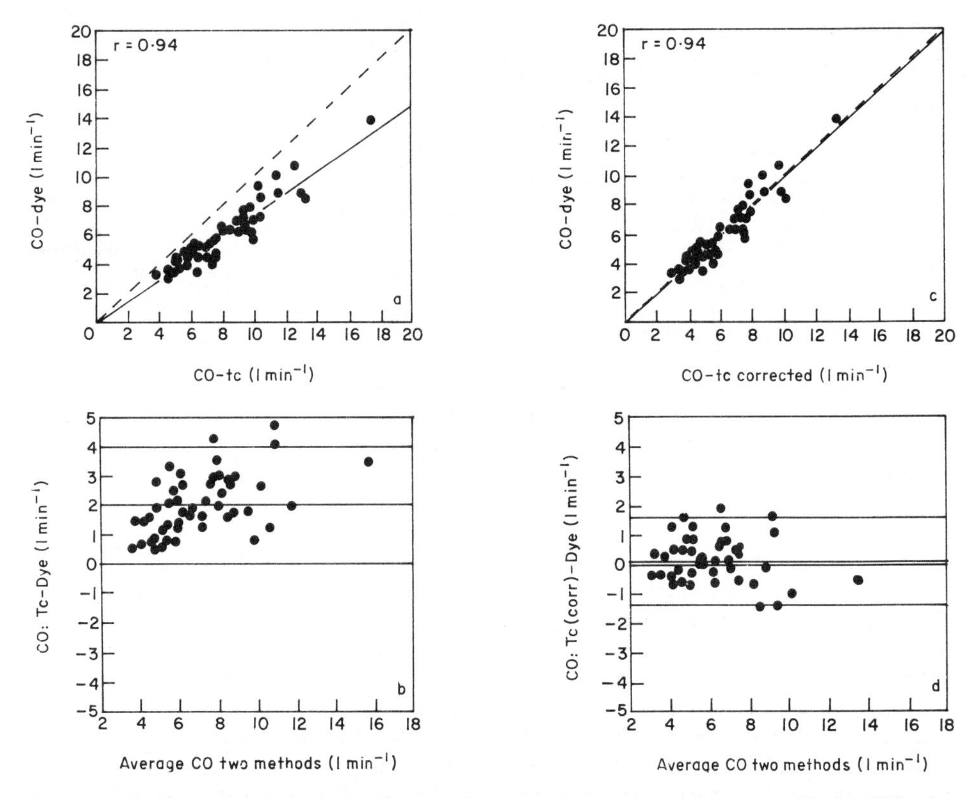

Figure 2 (a) Comparison between duplicate cardiac output measurements by isotope dilution (CO-tc) and dye dilution (CO-dye), with regression line and line of equality.
(b) Difference against mean for cardiac output data. The solid lines indicate mean difference and 95% confidence intervals.
(c) Comparison between duplicate cardiac output measurements by isotope dilution (CO-Tc corr) and dye dilution (CO-Dye) after statistical correction based on the regression equation CO-dye $= -0{\cdot}02 + 0{\cdot}75$ CO-Tc (panel a).
(d) Difference against mean for corrected cardiac output data. The solid lines indicate mean difference and 95% confidence intervals.

however, that reduction of counting of extracardiac activity by narrow-angle collimators may reduce the information obtained from the radiocardiogram. In addition it makes localization of the detector over the heart more critical. Using the statistical, empirical derived, correction factor, the systematic difference between the two methods was nullified (Fig. 2, panel c). Although 'statistical correction' thus reduced bias, it did not influence the distribution of the difference between the two methods, as delineated by the limits of agreement (Fig. 2, panel d).

PRECISION AND REPRODUCIBILITY

For a proper assessment of the ability of radiocardiography to detect statistically significant changes in cardiac output, one should take into account that the overall biologic error of any dilution method of cardiac output determination, including the standard dye dilution technique, is in the order of 10–20%[27]. Minute-to-minute physiologic variation in cardiac output and the imprecision of intermittent measurements together with factors such as technique of injection, adequacy of mixing and accuracy of 'curve sampling' play an important role. Therefore, the precision of any estimate of cardiac output is served by multiple measurements. However, in contrast to practice with the dye dilution and thermodilution techniques, serial measurements by the isotope dilution technique are difficult to perform. In particular, with the gamma camera radiation dose and a disturbingly high background activity preclude multiple injections. With the

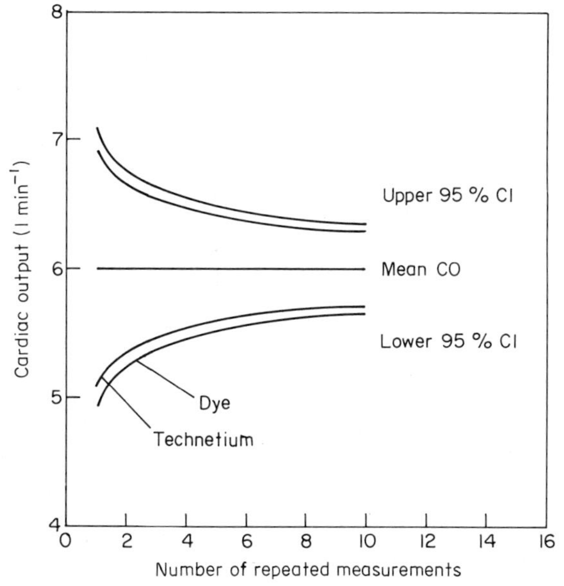

Figure 3 Relationship between the number of replicate measurements and the reliability (95% confidence interval) of isotope (technetium) or dye dilution cardiac output estimation.

probe, duplicate measurements are feasible and, in our practice, standard. The coefficient of variation of our series of duplicate measurements (N = 38) with the probe technique was 6% and with the dye technique 8%, which are similar to those reported in the literature. Does this mean that duplicate measurements are precise enough? We are unconvinced.

In this regard a more informative approach is one-way analysis of variance of replicate measurements[28]. By this analysis, one can calculate the 95% confidence interval of a single measurement. In our isotope series, for a cardiac output of 6 l min^{-1}, the 95% confidence interval of a single measurement ranged from 5·1 to 6·9 l min^{-1}. The 95% confidence interval of a single dye dilution cardiac output appeared to be in the same order of magnitude. Multiple measurements reduce the width of the 95% confidence interval. What this implies for our isotope and dye dilution data is shown in Fig. 3. Here, the hypothetical relationship between the number of replicate measurements and the reliability of estimation of cardiac output is shown. Both methods appeared to have comparable precision. The width of the 95% confidence interval is reduced markedly when two or more measurements are performed. Given the limited number of injections possible with the isotope technique, the unreliability of these measurements remains considerable.

Quantitative radiocardiography by the gamma camera

Several recent studies have emphasized the use of the gamma camera for determination of cardiac output. Of the two techniques currently employed, gated blood-pool imaging (determination of cardiac output based on estimates of ventricular volume and ejection fraction) and first-passage isotope dilution, each have theoretical and practical limitations. Difficulty in precise definition of ventricular edges and unreliable results in patients with valvular regurgitation and/or instable rhythms are the major limitations of the blood-pool imaging method.

According to Fouad *et al.*[14] first-passage isotope dilution with ^{99m}technetium albumin as tracer and an automated computer program for gamma camera data analysis allow a rapid and reproducible measurement of cardiac output. In their series, cardiac output determined by computer selection of either left ventricular area of interest or the whole heart correlated significantly with that determined by dye dilution ($r = 0{\cdot}86$, $P < 0{\cdot}001$, for both). However, analysing their actual data and plotting the difference between the methods against the mean (Fig. 4), the agreement between the two methods is not so clear. Although the source of the bias in this study cannot be precisely assessed, it is of sufficient magnitude that it cannot be ignored. Site and speed of bolus injection[29], type and in vivo instability of the bond of the carrier substance for the isotope[30] and extracardiac background radioactivity[21] all could have contributed. Kelbaek *et al.*[16] also developed a computer program for calculation of cardiac output. They used in vitro ^{99m}technetium-labelled red blood cells as tracer and followed by gamma camera the first passage of the bolus through the heart. For calculation of the area under the time-activity curve, they selected the left ventricle as area of interest and, in doing so, cardiac output was clearly overestimated compared with that obtained by thermodilution. They proposed that, even with appropriate selection of an area of interest, correction for background activity during equilibrium counting was mandatory. If their thesis were true, one of the major arguments for use of the gamma camera instead of the probe system would be invalid. However, the findings of Kelbaek *et al.* were at variance with those of Glass *et al.*[15] In a careful analysis, the latter group showed that selection of region of interest but not background correction yielded the best correlations with cardiac output by thermodilution. In a letter to the editor,

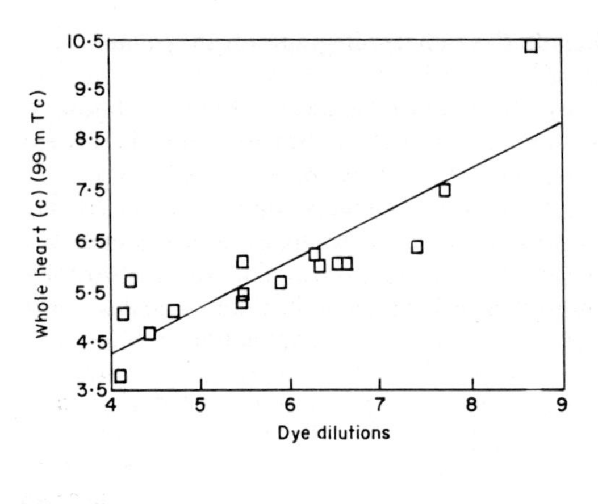

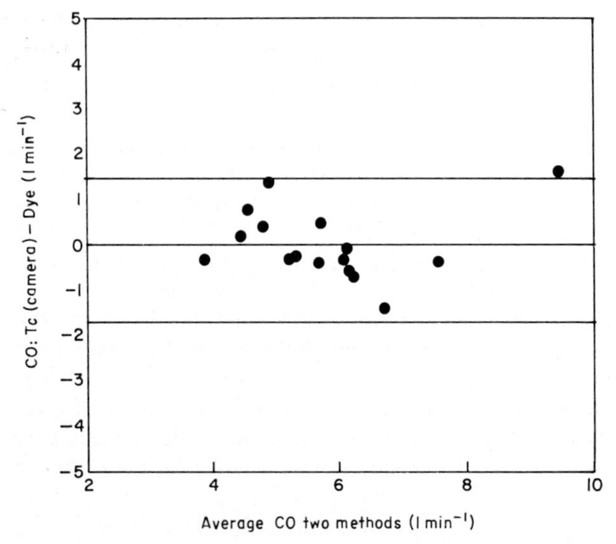

Figure 4 From the graph (upper panel) published in Fouad's paper[14], the mean difference between isotope (99mTC) and dye dilution measurements (lower panel) was calculated. The solid lines indicate mean difference and 95% confidence intervals.

they pointed out that discrepancies between the results of their study and that of Kelbaek *et al.* could be ascribed to a magnitude of factors rather than to background activity alone[17]. They concluded that as long as hardware (type of camera, collimator) and software (tracer, area of interest, curve analysis) are not standardized, each institution has to do its own evaluation. Probably they are right and, for the time being, without standardization of methods, the controversy about these issues will continue.

Conclusions

Bedside clinical examination is undoubtedly insufficient for a proper assessment of the adequacy of the function of the cardiovascular system. Therefore, since the turn of the century, both physiologists and clinicians have put a great deal of effort into the measurement of the output of the heart and the estimation of the resistance that the vasculature offers to flow. As a result, a wide variety of methods to measure cardiac output is available today. This short review focuses on the place and use of external counting techniques in studies of the circulation.

In general, assessment of global myocardial performance would be best served by gated equilibrium cardiac blood pool scintigraphy, because it enables simultaneous measurement of ventricular dimensions, ventricular wall motion, ejection fraction and volumetric cardiac output. However the available techniques of measurement are very complex and time consuming. In addition, their accuracy and reproducibility are still open to question and standardization is mandatory before these very complex computer techniques can be accepted as routine clinical tools.

In contrast, first-pass radiocardiography is relatively simple and allows a rapid estimation of cardiac output per se. However, because it is an indicator dilution method, it is prone to the same errors as the indocyanine green dye and thermodilution methods. As indicated above, radiocardiography by probe or gamma camera can be subject to even greater error. Our review of four recent papers[14–16,22] comparing radionuclide cardiac output with results obtained by green dye or thermodilution showed differences in individual patients ranging from +41 to −22%. Taking all data together, the 95% confidence limits of the mean differences were $\pm 1{\cdot}3\,l\,min^{-1}$, which represent ±24% of the mean invasive cardiac output ($5{\cdot}4\,l\,min^{-1}$). It should be stressed, however, that the overall biologic error of the invasive standard techniques is also considerable[27,31,32] and that the error of the radionuclide method is thus in the same 10–20% range.

The precision of estimation of cardiac output by any indicator dilution technique would be served by repetitive measurements. Unfortunately, the radiopharmaceutical used in radiocardiography, ^{99m}technetium, has characteristics that are less than ideal for repetitive studies within a short period of time. A half-life of 6 h, leading to disturbingly high background, limits serial studies to one or two injections when using a dose of 10–15 mCi. In this regard, the probe has some advantage over the gamma camera, but even with the probe, multiple injections are not feasible. Very recently, ^{191m}iridium has been proposed as an alternative to ^{99m}technetium[33]. ^{191m}Iridium has the desirable physical characteristics of a very short half-life (4·9 s) and dual peak energy levels (65 and 129 keV) that can be easily imaged by a conventional high count rate gamma camera. These characteristics would allow larger doses of radioactivity with significantly lower radiation exposure to the patient, a near absent background in serial studies and an end result of better counting statistics. However, as with any new isotope, the potential application of iridium are yet to be fully explored.

The final question that arises is whether a single isolated, not very precise, non-invasive estimation of cardiac output by radiocardiography is needed in clinical practice. The answer is no. The clinical utility of radiocardiography lies in long-term functional assessment of myocardial performance in hypertensive patients with cardiac disease who are undergoing pharmaceutical interventions. Provided technology becomes more standardized and accuracy is improved, this application may yield unique information about the circulation and the potential benefits of pharmacotherapy.

We gratefully acknowledge the help of Mrs Sylvia C. van Eijsden-Noordermeer, who assisted with the preparation of the manuscript.

References

[1] Hamilton WF, Moore JW, Kinsman JM, Sparling RG. Simultaneous determinations of the greater and lesser circulation times, of the mean velocity of blood flow through the heart and lungs, of the cardiac output and an approximation of the amount of blood activity circulating in the heart and lungs. Am J Physiol 1928; 85: 377.

[2] Fox IJ. History and developmental aspects of the indicator-dilution technique. Circ Res 1962; 10: 381–92.

[3] Prinzmetal M, Corday E, Spritzler RJ, Flieg W. Radiocardiography and its clinical applications. Am J Med 1949; 139: 617–22.

[4] Donato L, Rochester DF, Lewis ML, Durand J, Parker JO, Harvey RM. Quantitative radiocardiography. II. Technique and analysis of curves. Circulation 1962; 26: 183–8.

[5] Pritchard WH, McIntyre WJ, Moir TW. The determination of cardiac output by the dilution method without arterial sampling. II. Validation of precordial counting. Circulation 1958; 18: 1147–55.

[6] Carter BL, Johnson SE, Loeffler RK *et al.* A critical evaluation of external body surface counting in the determination of cardiac output with radio-active isotopes. Am J Roentgenol 1959; 82: 618–25.

[7] Gorten RJ, Gunnels JC. Disclosure of a systematic difference in cardiac output values determined by external counting and arterial sampling technics. Circulation 1960; 22: 757–8.

[8] Conn HL. Use of external counting technics in studies of the circulation. Circ Res 1962; 10: 505–17.

[9] Boudreau RJ, Loken MK. Functional imaging of the heart. Semin Nucl Med 1987; 17: 28–38.

[10] Slutsky R, Karliner J, Ricci D *et al.* Left ventricular volumes by gated equilibrium radionuclide angiography: a new method. Circulation 1979; 60: 556–64.

[11] Sorensen SG, Ritchie JL, Caldwell JH, Hamilton GW, Kennedy JW. Serial exercise radionuclide angiography. Validation of count-derived changes in cardiac output and quantitation of maximal exercise ventricular volume change after nitroglycerin and propranolol in normal man. Circulation 1980; 61: 600–9.

[12] Dehmer GJ, Firth BG, Lewis SE, Willerson JW, Hillis LD. Direct measurement of cardiac output by gated equilibrium blood pool scintigraphy: validation of scintigraphic volume measurements by a nongeometric technique. Am J Cardiol 1981; 47: 1061–6.

[13] Massie BM, Kramer BL, Gertz EW *et al.* Radionuclide measurement of left ventricular volume: comparison of geometric and count-based methods. Circulation 1982; 65: 725–30.
[14] Fouad FM, Houser T, MacIntyre WJ, Cook SA, Tarazi RC. Automated computer program for radionuclide cardiac output determination. J Nucl Med 1979; 20: 1301–7.
[15] Glass EC, Rahimian J, Hines HH. Effect of region of interest selection on first-pass radionuclide cardiac output determination. J Nucl Med 1986; 27: 1282–92.
[16] Kelbaek H, Heslet L, Skagen K, Munck O, Godtfredsen J. First passage radionuclide cardiography for determination of cardiac output: evaluation of an improved method. J Nucl Med 1987; 28: 1330–4.
[17] Glass EC. First-pass radionuclide determination of cardiac output: an improved gamma camera method. J Nucl Med 1988; 29: 1154–5.
[18] Kelbaek H, Harling OJ, Munck O. First pass radionuclide determination of cardiac output: an improved gamma camera method. J Nucl Med 1988; 29: 1155.
[19] Donato L, Giutini C, Lewis ML *et al.* Quantitative radiocardiography. I. Theoretical considerations. Circulation 1962; 26: 174–82.
[20] McIntyre WJ, Pritchard WH, Moir TW. The determination of cardiac output by the dilution method without arterial sampling. I. Analytical concepts. Circulation 1958; 18: 1139–46.
[21] Kelbaek H, Gjorup T, Hartling OJ *et al.* The influence of a background correction that considers the heart volume on radionuclide left ventricular ejection fraction determinations. Br J Radiol 1986; 59: 993–6.
[22] Man in 't Veld AJ, Wenting GJ, Verhoeven RP, Schalekamp MADH. Quantitative radiocardiography by single-probe counting using ^{99m}technetium-albumin. Clinical application in follow-up studies. Neth J Med 1978; 21: 166–75.
[23] Bland JM, Altman DG. Statistical methods for assessing agreement between two methods of clinical measurement. Lancet 1986; i: 307–10.
[24] Van der Feer Y, Douma JH, Klip W. Cardiac output measurement by the injection method without arterial sampling. Am Heart J 1958; 56: 642–51.
[25] Shackman R. Radio-active isotope measurement of cardiac output. Clin Sci 1958; 17: 317–25.
[26] Steele PP, Van Dyke D, Trow RS, Anger HO, Daries H. Simple and safe bedside method for serial measurement of left ventricular ejection fraction, cardiac output, and pulmonary blood volume. Br Heart J 1974; 36: 122–31.
[27] Levett JM, Repogle RL. Thermodilution cardiac output: a critical analysis and review of the literature. J Surg Res 1979; 27: 392–404.
[28] Fleiss JL. Reliability of measurement. In: Fleiss JL, Ed. The design and analysis of clinical experiments. New York: Wiley Medical, 1986: 1–15.
[29] Fouad FM, Tarazi RC, McIntyre WJ, Durant D. Venous delay, a major source of error in isotopic cardiac output determination. Am Heart J 1979; 97: 477–84.
[30] Millar AM, Hannan WJ, Sapru RT *et al.* An evaluation of six kits of technetium-99m human serum albumin injection for cardiac blood pool imaging. Eur J Nucl Med 1979; 4: 91–4.
[31] Sleeper JC, Thompson HK, McIntosh HD, Elston RC. Reproducibility of results obtained with indicator-dilution technique for estimating cardiac output in man. Circ Res 1962; 21: 712–20.
[32] Hillis LD, Firth BG, Winniford MD. Analysis of factors affecting the variability of Fick versus indicator dilution measurements of cardiac output. Am J Cardiol 1985; 56: 764–8.
[33] Hellman C, Zafir N, Shimoni A, Issachar D *et al.* Evaluation of ventricular function with first-pass iridium 191m radionuclide angiocardiography. J Nucl Med 1989; 30: 450–7.

Doppler ultrasonic measurement of cardiac output: reproducibility and validation

A. J. S. Coats

Department of Cardiovascular Medicine, John Radcliffe Hospital, Oxford, U.K.

KEY WORDS: Haemodynamics, Doppler ultrasound, reproducibility, validation.

The Doppler ultrasonic estimation of cardiac output in man is reviewed. Minimal requirements for accurate measurements are discussed, and the published results of reproducibility studies and validation studies are summarized and analysed. Analysis of Doppler records has a coefficient of repeat determination of 5–8% for aortic or LV outflow tract measurements and this is higher for other sites. Short-term variability varies from 4 to 10%, and that over days to weeks from 9 to 14%. Thus a single measurement may vary up to ±28% over time with no true change in cardiac output.

For cardiac output determination, the Doppler methods show accuracies varying from 10 to 22% (coefficient of variation of the differences between methods) indicating that a single aortic based measurement only reliably lies within ±28% compared with other 'standard' methods, and during exercise the accuracy is less (±44%).

Doppler methods are safe, fairly reproducible and reasonably accurate methods for measuring cardiac output in selected patients provided signal quality is adequate during recording.

Introduction

Doppler ultrasound has been used to measure cardiac output in a variety of different ways all based on the measurement of blood velocity at some point in the circulation. This group of methods has been extensively discussed and used in the literature over the last few years[1–6]. The Doppler methods constitute useful non-invasive techniques of reasonable accuracy in experienced hands. There are, however, important restrictions on their applicability to clinical and research purposes.

Physics

The physical principles underlying the use of the Doppler shift to measure blood velocity have been well described[7–9]. When an ultrasonic wave meets a moving object, any reflected wave so produced will have a frequency shift proportional to the reflecting object's velocity vector in the direction of the incident wave. Thus the measurement of the frequency shift produced by ultrasound reflecting off a moving column of blood can be used to measure the velocity vector of the blood parallel to the ultrasound beam. More detailed descriptions of the equations involved have been published[1]. From a knowledge of blood velocity in the aorta, and other parts of the circulation, estimates can be made of blood flow if the cross-sectional area of the moving column of blood is known. More detailed descriptions of the validity of assumptions necessary for such calculations are given below.

In evaluating the Doppler method for measuring cardiac output, it is necessary to discuss the theoretical limitations and assumptions inherent in this technique before addressing comparative data on this technique vs other direct and indirect measurements of blood flow. The reproducibility of Doppler recordings is also reviewed.

Address for correspondence: Dr A. J. S. Coats, Department of Cardiovascular Medicine, John Radcliffe Hospital, Oxford, OX3 9DU, U.K.

Methods

SITE OF RECORDING

Measurements of cardiac output have been made at many points in the circulation, including the tricuspid valve[10], pulmonary artery, mitral valve, left ventricular outflow tract, aortic valve, ascending aorta[2], aortic arch and descending aorta[11]. Each site has particular advantages and disadvantages taking into account such factors as (i) ability to obtain insonation parallel to blood flow, (ii) the profile of the blood velocity distribution across the blood flow, (iii) the proportion of

Table 1 Measurement sites for cardiac output by Doppler and their associated characteristics

Site	Proportion of cardiac output	Velocity profile	Angle	CSA shape and size	Echo image quality	Signal to noise ratio
Tricuspid	Full pulmonary	Flat	Not parallel	Complex, variable	Poor	Poor
Pulmonary	Full pulmonary	Flat	Not parallel	Circular, stable	Poor	Poor
Mitral	Full systemic	Flat	Parallel	Complex, variable	Good	Poor
LVOT	Full systemic	Flat	Parallel	Circular, stable	Good	Good
Aortic Annulus	Full systemic	Flat	Parallel	Circular, stable	Good	Good
Asc. Aorta	95% systemic	Flat	Parallel	Circular, stable	Good	Good
Aortic Arch	75% systemic	Skewed	Parallel	Circular, stable	Good	Good
Desc. Aorta	75% systemic	Parabolic	Not parallel	Circular, stable	Poor	Good

CSA = cross-sectional area, LVOT = left ventricular outflow tract.

the total cardiac output traversing the vessel or valve, (iv) the velocity signal to 'noise' ratio and (v) the quality of echocardiographic imaging of the whole of the vessel or valve circumference to measure cross-sectional area. The features specific to each measurement site are listed in Table 1.

INSONATION ANGLE

If it proves impossible to align insonation parallel with blood flow then corrections can be made if the angle is known, but this will increase error in calculating flow and is not recommended. The further the angle from parallel the more serious will be the effect of minor inaccuracies of angle. Parallel insonation is usually achieved for the left ventricular outflow tract from the apical approach, for the aortic valve or ascending aorta from the suprasternal approach or for the mitral valve from the apical approach. It is not easy for pulmonary artery or tricuspid valve studies.

VELOCITY DISTRIBUTION PROFILE

The calculation of blood flow from velocity integral multiplied by cross-sectional area depends on the assumption that the velocity profile at the measurement site is uniform across a vessel or chamber. This is reasonable for sites that approximate to inlet conditions where blood is accelerating through a narrowing[12]. These conditions exist at most valve inlets and immediately downstream of a valve, but are lost as the blood moves further down the vessel; for example, in the aorta the blood velocity profile becomes initially skewed and then parabolic in profile as flow progresses around the aortic arch[13–15].

PROPORTION OF CARDIAC OUTPUT

As the Doppler methods measure blood velocity at one point in the circulation they only measure volume flow at that point, and hence only measure cardiac output if that point receives the full cardiac output. The term 'cardiac output' usually refers to left ventricular output which enters and stays in the arterial system. Thus in aortic regurgitation more blood is ejected into the ascending aorta than travels down the aorta and hence Doppler aortic valve methods will overestimate net cardiac output[16]. Similarly in arteriovenous shunting or intracardiac shunts there may be discrepancy between right and left ventricular outputs[17]. Some measurement sites receive only a proportion of the total output. For example, a measurement taken several centimetres down the aorta from the valve will not include coronary blood flow and hence will be approximately 5% less than total cardiac output; nevertheless this will be a fairly consistent error even during exercise. Descending aortic measurements do not include all the flow to the heart as well as the head and neck vessels (approximately 25% of total cardiac output) and the proportion may not be stable with changing physiological states.

VELOCITY SIGNAL TO NOISE RATIO

Doppler techniques cannot detect very low blood velocities because of the noise produced by structural echoes at near-zero frequency shifts. Thus there is always the need for high-pass filtering in Doppler signals. This filtering should be as minimal as possible, while excluding the low velocity noise. The best signal to noise ratios are, therefore, found at areas in the circulation with the highest blood velocities. For this reason, the aortic and left ventricular outflow tract methods have an advantage over the lower velocities of the pulmonary artery and the very much lower velocities across the pulmonary and mitral valves.

CROSS-SECTIONAL AREA MEASUREMENTS

The cross-sectional area of the vessel, valve or heart chamber must be measurable accurately at the level blood velocity measurements are taken. Poor image quality often limits measurement of pulmonary artery cross-sectional area because the whole circumference of the vessel wall may not be imaged[2,18]. Similar limitations exist for the tricuspid valve[10].

Any assumptions about the geometrical shape of the chamber may be important in calculating cross-sectional area from dimensional measurements of wall separations. The mitral valve annulus for example is approximately ellipsoid, rather than circular, and the geometrical shape may vary between patients and in different physiological conditions so accurate determination of shape may be important in calculating cross-sectional area and, hence, cardiac output[19,20]. For the mitral valve methods, a more comprehensive estimate of annular area corrected for variable shape and size during diastole is superior to assumption of simple circular shape both in theory and in practice[19,20].

During exercise and drug-induced changes in cardiac output cross-sectional area seems to change substantially at the mitral valve level[21], and it is also likely to do so at the tricuspid level. At the aortic annulus there is little or no change in cross-sectional area with exercise[21], and a maximal change of only 10% at large increases in cardiac output[22]. It would appear, therefore, that separate estimates of cross-sectional area with each measurement are necessary for mitral and tricuspid estimates, but are probably not essential for aortic methods.

When only percentage changes in cardiac output are required and all comparisons are within a single subject, the use of stroke distance (velocity integral) and minute distance (stroke distance times heart rate) instead of stroke volume and cardiac output is supportable. When comparing single measurements from different subjects, however, it is important to realize that any error in estimating aortic diameter will be squared and hence imperceptible differences in aortic root diameter might have substantial effects on cardiac output calculations. It is advisable, therefore, to consider the Doppler method primarily suitable for measuring changes in cardiac output within a subject rather than comparing the absolute values of cardiac output between different subjects.

ECHOCARDIOGRAPHIC TECHNIQUES

The technique depends on the measurement site chosen (Table 1). The site measured should be exactly where the Doppler measurement is made. The most feasible technique for routine use is echocardiography — A-mode, M-mode, or two-dimensional. Radiographic, magnetic resonance imaging and direct measurement at the time of surgery or at post-mortem have been used but have obvious impracticalities. The use of nomograms for estimating aortic cross-sectional area from age, sex, height, weight etc. have not been validated and cannot be recommended[23]. The use of A-mode or M-mode echo with measurement of diameter depends upon calculations based on certain shape assumptions, and usually involving squaring a diameter measurement (or multiplication of two different measurements together) thereby exaggerating any variability in the diameter measurement. Alternatively, the use of two-dimensional echo with planimetry of the vessel or valve apparatus area should theoretically be better, but if the plane of the image is not perpendicular to the vessel the area so measured will be inaccurate. Whether the plane is at an angle may be very difficult to determine.

There is no consensus on the details of echocardiographic measurements at any of the sites used for Doppler measurements as different studies have suggested different guidelines for the most accurate results. Even for a single site the results are not consistent. Some Doppler studies find the use of the smaller cross-sectional area measured by the trailing edge to leading edge method gives better correlations with other methods[24]. Although in theory this echocardiographic convention measures internal diameter, it is known that variable ultrasound intensity will produce variable echo persistence and hence tend to reduce the measured diameter as the gain of the apparatus is increased. In contrast, other workers have achieved their best results using the leading edge to leading edge convention which might slightly overestimate vessel diameter[25]. The discrepancy probably relates to differences in Doppler methodology.

In summary, taking into account all the features on choice of site and echocardiographic measurement of cross-sectional area, the aortic annulus and ascending aortic measurements appear the most reliable and simplest to determine.

Table 2 Clinical features which produce inaccurate or unreliable Doppler estimates of cardiac output

Absolutely unreliable or impossible	Relatively unreliable
Valvular stenosis or regurgitation*	Murmur at measurement site
Prosthetic heart valve*	Atrial fibrillation
Gross bodily movement	Other arrhythmia
Low blood velocity	Difficult ultrasound access
	Obstructive airway disease
	Chest movements, e.g. heavy exercise

* = At the site of measurement.

TYPE OF ULTRASOUND

The two main methods are continuous wave (CW), with no limit to detectable frequency but no depth resolution, and pulsed wave (PW), with depth resolution but a maximal detectable velocity depth-product and hence maximal detectable peak velocity[1]. Although similar results have been reported on accuracy vs other methods and reproducibility, there are theoretical advantages in favour of pulsed wave; that is, the position of the recording is known and can be fixed, so repeat measurements will always be taken at approximately the same point in the circulation. This is crucial for complex sites such as the mitral, tricuspid and pulmonary valves. Also full spectral display gives a clearer indication of signal quality with PW since it is obvious by the absence of vessel wall artefacts whether the sample volume lies entirely within the bloodstream for the whole of the cardiac cycle. If so, it is much more likely that the ultrasound beam is nearly parallel to the blood flow. Pulsed wave is also more likely to give the true average blood velocity at the point of sampling because the frequency data can be analysed to give the intensity-weighted mean frequency, so the method does not rely on estimates of modal or even maximal detectable velocities which do not reliably reflect the spatially averaged blood velocity. The limitation on peak velocity detected by PW is not a major drawback in adults because blood velocities above 150 cm s^{-1} are rare in the absence of valvular disease or shunts, so the velocity limit is rarely reached, even during exercise. Procedures exist for further increasing the maximal detectable velocity by PW Doppler if necessary.

CLINICAL APPLICABILITY

Patient selection

Estimates of the proportion of subjects suitable for Doppler studies vary from 50 to nearly 100%, depending on the age and clinical status of the subject, and on the Doppler method used. Inaccurate measurements are inevitable in certain conditions and are likely in some others (Table 2). In addition, certain measurement sites have particular problems, such as short neck, unfolded or dilated aorta for the suprasternal approach, obstructive airway disease for the apical approach, right to left shunt for the tricuspid valve or pulmonary artery methods, and atrial fibrillation for the mitral or tricuspid methods. In any situation it is advisable to have a full spectral display of the returning frequency shifts so that the operator is aware of signal quality to guide recording and to show when record quality is inadequate. Hence, the use of devices with only rudimentary analogue displays or digital outputs is inadequate because they do not permit signal quality to be determined.

On the whole, good quality signals are usually obtainable in the majority of children and young adults, and in many patients, the most reliable measurement sites being the apical approach for the left ventricular outflow tract or the suprasternal approach for the ascending aorta or aortic valve[26].

Clinical uses

As it is possible to measure Doppler-derived flow at several points in the circulation, it is possible further to compare the flows across two valves and hence estimate regurgitant fractions[16] or the size of shunts[17,27], in addition to measuring cardiac output, vascular resistance and their responses to treatment or interventions.

REPRODUCIBILITY

Reproducibility is essentially how reliably a single measurement can predict subsequent measurements. Therefore, it includes both the variability of the method itself and the biological variability of what is being measured. This information is essential as a guide to judging the importance of changes in measurements. If a detected change in cardiac output is less than the inherent variability of the method, then one can have no confidence that a real change has taken place. In a controlled trial, statistical evaluation of cardiac output measurements in the treatment vs placebo groups in effect compares treatment-related effects with background variability; clearly, a poorly reproducible method will be inefficient for detecting real changes at statistical significance. Some measurements of reproducibility are essential before statistical power calculations to determine trial size can be performed for a proposed study.

Reproducibility can be measured in several ways, and can be directed at various aspects of a measurement technique. Mathematically the reproducibility of a measurement x_i can be expressed as: (1) the standard deviation or variation coefficient of all measurements x_i to x_n in a single subject; (2) the mean differences between two measurements x_i and x_{ii}; (3) the standard deviation or variation coefficient of the differences between two measurements x_i and x_{ii}; (4) an analysis of the components of variance on multiple measurements in a large group of subjects to define the proportion of variability attributable to such factors as method, patient, operator, analyser and time, as well as the interactions between these factors; or (5) a correlation analysis of the relationship between paired measurements x_i and x_{ii}. (This last method is poor as it depends strongly on the range of measurements and not on the reliability of measurements[28].)

These variabilities need to be known for both the short-term and long-term as changes in cardiac output may be investigated over either time course. Reproducibility needs ideally to be investigated in the type of patients the method will be used for (not always on healthy young volunteers) and over the time-course of interest (e.g. up to months in a drug trial). Method 3 (above) is the best to assess the usefulness of a method for a clinical trial since measurements can be repeated after a typical treatment period in a typical population for study with no intervention or placebo, and the results, expressed as standard deviations of differences between the repeat measurements, can be used to calculate trial power using published nomograms for parallel group[29] or crossover design trials[30]. Method 2 is the most commonly quoted but is difficult to assess because the results are usually quoted as mean percentage differences between paired readings, a typical result for cardiac output over the short-term being 6·8% ± 5·0%[16]. It is obvious from this that the distribution of mean percentage difference is skewed because two standard deviations below the mean difference would be −3·2% and it is impossible to have less than no difference between methods (the mean percentage difference is an absolute value which cannot be negative). Therefore the standard deviation of 5·0% with a mean of 6·8% probably means the absolute differences range to well over 20%, and 6·8% is an overly optimistic summary of the reproducibility, unsuitable for statistical power calculations.

Also of importance in the interpretation of reproducibility is the design of the study in which the measurements were taken. The extremely good reproducibility figures sometimes quoted almost certainly suggest that the measurements were not taken in a typical patient sample during the rigours of a clinical trial.

In Table 3 are listed studies which describe the reproducibility of Doppler estimates. Where only a single measurement of conduit cross-sectional area was made, the reproducibility is reported for stroke distance or minute distance rather than stroke volume or cardiac output, because variability in the cross-sectional area measurement has not been taken into account. Unfortunately, few studies report repeated independent and blinded measurements separated by a period similar to that over which an intervention may be assessed.

As a summary of reproducibility of Doppler estimates of cardiac output, taking those studies which have analysed reproducibility according to Method 3 (above), the standard deviation of differences between repeat analyses of the same record is approximately 5–8% for aortic methods and 6–20% for the other methods. The short-term (minutes to hours) reproducibility of cardiac output measurements in total is approximately 4–10% for aortic and mitral valve methods. The medium-term reproducibility (days to weeks) is approximately 9–14% for aortic methods. The measurements appear to be slightly less reproducible during exercise than at rest (Table 4). These figures and published nomograms[29,30] suggest that aortic Doppler recordings would have a 90% chance of detecting a 7% change in cardiac output over minutes to hours or a 12% change over days to weeks in 10 subjects in a crossover trial or 40 subjects in a parallel group trial.

VALIDATION STUDIES

There is no established standard for the measurement of cardiac output in man, and hence the only

Table 3 Short-term (minutes to hours) and medium-term (days to weeks) reproducibility studies of Doppler cardiac output or stroke volume

First author	Subjects	Technique	Short-term	Medium-term	Analyser
Fraser[34]	21 N	CW DA	6·0% (SD)[a]		
Gisvold[35]	5 N	PW AA		11·7% (SD)[a]	
				14·2% (MD)[a]	
Lewis[16]	15 P	PW MV	16·4% (SV)[b]		
		PW LVOT	6·8% (SV)[b]		
Ihlen[36]	10 P	PW AA		9·2% (SD)[a,c]	
Fagard[37]	10 P	PW AA	5·8% (SD)[a,d]		
			6·9% (MD)[a,d]		
Mehta[38]	40 N	CW AA		10·8% (SD)[a,c]	
Robson[39]	8 N	CW AA	6·4% (SD)[a]		
			7·8% (SV)[a]		
			8·8% (CO)[a]		
DeZuttere[40]	30 P	PW MV			8·4% (SD)[b]
Robson[41]	14 N	CW AA	4·8% (CO)[b]		
		PW MV	4·1% (CO)[b]		
		PW PV	4·1% (CO)[b]		
Hoit[42]	10 P	PW MV			5·1% (CO)[b]
Looyenga[33]	50 P	DBPW AA	9·9% (CO)[b]		
Kumar[31]	14 P	CW oes	10% (CO)[b]		
Chandraratna[43]	18 N	CW AA	0·97 (CO)[f]		
	29 N	CW AA	0·98 (CO)[g]		
Gardin[44]	10 N	PW AA		3·8% (SD)[b,h]	
Nicolosi[45]	30 N	PW MV			6·1% (CO)[a]
McLennan[46]	142 N	CW DA	5·3% (SD)[b]		5·5% (SD)[b]
			5·4% (MD)[b]		5·7% (MD)[b]
Nicolosi[47]	30 N	PW AA			7·8% (CO)[a]
		PW LV			7·4% (CO)[a]
		PW PA			7·9% (CO)[a]
		PW MV	(corrected for leaflet separation)		7·7% (CO)[a]
		PW MV	(circular assumption)		20% (CO)[a]
		PW TV			12% (CO)[a]

N = normal subjects, P = patients, PW = pulsed wave, CW = continuous wave, DBPW = dual beam pulsed wave, AA = ascending aorta, LV = left ventricular outflow tract, MV = mitral valve method, PA = pulmonary artery, oes = oesophageal probe, SD = stroke distance, MD = minute distance, SV = stroke volume, CO = cardiac output. a = coefficient of variation, b = mean percentage difference, c = all readings done at the same depth, therefore not entirely blind, d = readings were not taken blind to previous recordings, e = note that the same study reported previously in abstract form[48] gives very much smaller values (4·9%) for apparently the same design, patient numbers and statistical analysis, f = correlation analysis between two readings by same observer, g = correlation analysis between two readings by two observers, h = the results compare the average value of two analysers for each record therefore much of the variability has been averaged out in calculating this result.

Table 4 Reproducibility during exercise

First author	Subjects	Technique	Short-term	Medium-term
Ihlen[49]	10 P	PW AA	6·2% (SD)[a]	
Mehta[38]	40 N	CW AA		11–12% (SD)[b]

Abbreviations as for Table 3. a = within-subject coefficient of variation of measurements taken by two observers and on 2 days during upright 50W bicycle exercise. Therefore this averages short-term, long-term and inter-recorder variabilities, b = coefficient of variation.

Table 5 Validation studies of aortic Doppler cardiac output measurements

First author	No.	Technique	Standard	Timing	Range	r	slope	SEE or SDD	Analysis
Magnin[50]	11	PW AA PS	Fick	$\leqslant$24 h	2·1	0·83	1·67	N/S	Analogue
Alverson[51]	33 P	PW AA SS	Fick	N/S	13·9	0·98	1·07	0·22 (12%)	Analogue
Goldberg[18]	14 P	PW AA SS	Ind.	N/S	8	0·91	0·76	0·60 (15%)	Plan
Sanders[17]	27 P	PW AA Ap	Fick	N/S	4	0·78	1·23	0·81 (16%)	FFT Plan.
Huntsman[52]	45	CW AA SS	TD	N/S	5·5	0·94	0·94	0·58 (10%)	Analogue
Chandraratna[43]	36	CW AA SS	TD	Sim.	5·5	0·97	0·93	0·42 (8%)	Analogue
Ihlen[53]	10	PW AA SS	TD	Sim.	3·4	0·96	0·81	0·70 (12%)	C-Z
Ihlen[53]	11	PW AA SS	Fick	Sim.	2·7	0·94	0·82	0·70 (14%)	C-Z
Lewis[16]	35	PW LV Ap	TD	mins	4·5	0·91	1·18	0·63 (11%)	FFT Modal
Nishimura[54]	38	CW AA SS	TD	N/S	5	0·94	1·0	0·78 (15%)	Max.
Loeppky[55]	15	PW AA SS	Fick	Sim.	2·2	0·84	0·84	0·61 (14%)	ZCD
Rose[56]	16	CW AA SS	TD	Sim.	2·6	0·92	1·04	0·48 (11%)	N/S
Labovitz[57]	35	PW AA SS	TD	mins	5·1	0·85	N/S	0·99 (20%)	C-Z Modal
Labovitz[57]	23	PW DA SS	TD	mins	5·6	0·83	N/S	1·10 (21%)	C-Z Modal
Labovitz[57]	31	PW LV Ap	TD	mins	5·3	0·90	N/S	0·95 (19%)	C-Z Modal
Mehta[58]	21	CW AA SS	TD	Sim.	20·0	0·97	0·96	0·49 (10%)	Analogue
Vandenbogaerde[59]	28	CW AA SS	TD	mins	N/S	0·97	N/S	N/S	Analogue
Ihlen[49]	20	PW AA SS	TD	Sim.				(17%)	Analogue
Bojanowski[8]	12	PW AA SS	TD	Sim.	6·5	0·88	0·87	N/S	FFT IWM
Innes[60]	14	PW AA SS	TD	Sim.	4·3	0·92	0·90	0·90 (15%)	FFT IWM
Fagard[37]	10	PW AA SS	Fick	mins				1·15 (15%)	FFT Modal
Morrow[61]	12 P	CW AA SS	TD	Sim.	6	0·94	1·02	0·45 (13%)	Max.
Maeda[62]	34	CW AA SS	TD	Sim.	3·5	0·85	1·37	0·59 (14%)	Max.
Looyenga[33]	71	DBPW AA SS	TD	mins	7·4	0·94	0·94	0·62 (10%)	N/S

Timing = delay between the two measurements, range = range of standard method cardiac outputs in the study, slope = slope of the Doppler on standard method regression line, SEE = standard error of the mean, SDD = standard deviation of the differences, as explained in the text, and expressed as percentages in parentheses, P = paediatric, PW = pulsed wave, CW = continuous wave, DBPW = dual beam pulsed wave, AA = ascending aorta, LV = left ventricular outflow tract, DA = descending aorta, PS = parasternal approach, SS = suprasternal approach, Ap = apical or subxiphoid approach, TD = thermodilution, Ind. = indicator dilution, SEE = standard error of the estimate, SDD = standard deviation of the differences between the two methods, Plan = planimetry, Modal = modal velocity envelope, Max = maximal velocity envelope, IWM = intensity-weight mean velocity envelope, FFT = fast Fourier transform, ZCD = zero crossing detector, C-Z = Chirp-Z analyser, N/S = not stated.

Table 6 Validation studies of mitral valve, pulmonary artery, tricuspid valve, and oesophageal probe Doppler cardiac output

First author	No.	Technique	Method	Timing	Range	r	slope	SEE or SDD	Analysis
Goldberg[18]	14 P	PW PA PS	Ind.	N/S	8	0·72	1·11	1·11 (25%)	Plan
Sanders[17]	24 P	PW PA PS	Fick	N/S	7·7	0·88	1·15	2·4 (24%)	FFT Plan.
Labovitz[57]	35	PW PA PS	TD	mins	4·1	0·81	N/S	0·82 (16%)	C-Z Modal
Lewis[16]	35	PW MV Ap	TD	mins	4·5	0·87	1·25	0·59 (10%)	FFT Modal
Zhang[63]	20	PW MV Ap	TD	≤24 h	2·2	0·89	0·9	0·71 (11%)	Analogue
DeZuttere[40]	30	PW MV Ap	TD	mins	2·7	0·91	0·92	0·53 (9%)	Modal
Hoit[42]	48	PW MV Ap	TD	≤2 h	2·7	0·93	1·1	0·36 (9%)	Modal
Meijboom[10]	10 P	PW TV Ap	TD	mins	8	0·98	1·04	0·43 (10%)	FFT Modal
Kumar[31]	14	CW DA oes	TD	Sim.	16·5	0·76	0·93	1·76 (28%)	?

P = paediatric, PW = pulsed wave, PA = pulmonary artery, MV = mitral valve, TV = tricuspid valve, DA = descending aorta, PS = parasternal approach, Ap = apical approach, oes = oesophageal probe, TD = thermodilution, Ind. = indicator dilution, SEE = standard error of the estimate, SDD = standard deviation of the differences between the two methods, Plan = planimetry, Modal = modal velocity envelope, Max = maximal velocity envelope, IWM = intensity-weight mean velocity envelope, FFT = fast Fourier transform, ZCD = zero crossing detector, C-Z = Chirp-Z analyser.

Table 7 Validation studies of aortic Doppler velocity integral against electromagnetic velocity probes

First author	No.	Technique	Timing	Range	r	slope	SEE or SDD	Analysis
Innes[60]	5	PW AA SS	Sim.	N/S	0·89	0·97	N/S	FFT IWM
Mehta[38]	8	CW AA SS	Sim.	5	0·97	0·90	1·1 (9%)	Analogue
Coats[64]	6	PW AA SS	Sim.	4·1	0·99	1·0	0·34 (4%)	FFT IWM

Abbreviations as for Table 5.

way to test the accuracy of a new method is to compare it under a variety of conditions with older accepted methods even though the latter may be poorly reproducible or inaccurate on occasion. Thus for clinical validation of Doppler methods, workers have compared them with thermodilution, indicator dilution, direct and indirect Fick, radionuclide and radiographic ventriculography, electromagnetic catheters, echocardiography and electrical bio-impedance methods. The first three are those most commonly accepted as reliable in experienced hands.

In assessing a validation study, it is important to check that the subjects tested are representative of the patient population eventually to be studied with the new technique, both in terms of clinical status and physiological state (e.g. exercise). For example, validation in the paediatric population is inadequate for reliability testing in the adult. Also statistical tests used to compare methods are often inappropriate, as the most popular test (linear regression) addresses correlations between values rather than the accuracy of one method vs another. Bland and Altman have recently made convincing recommendations for the calculations of the standard deviation of differences (SDD) between two methods as the best way to validate a new method against an established one[28]. It is often not realized that SDD is virtually the same as the standard error of the estimate (SEE) derived from linear regression analysis of two methods. Unfortunately, many older studies do not report the SEE, only the r value. Summarized in Tables 5–8 are the data on validation studies for the Doppler methods at rest and during exercise, giving as much detail on SDD or SEE as can be gleaned from the papers reviewed.

The aortic Doppler methods correlate quite well with other standards with r values from 0·78 to 0·98 (mean: 0·91 ± 0·06), and SEEs from 0·22 to 1·15 (mean: 0·69 ± 0·24) $l\,min^{-1}$ and SDD as percentages from 8–22% (mean: 13·9 ± 3·4%). This suggests that for single measurements of cardiac output the aortic Doppler estimate has a 95% chance of lying within ±28% or 1·4 $l\,min^{-1}$ of the standard. There is no significant difference on average between the reported accuracy of CW and PW Doppler methods compared with other standard methods.

For the mitral method, results are not significantly different to those of the aortic method, r values being 0·90 ± 0·03, SEE 0·55 ± 0·15 $l\,min^{-1}$ and SDD (as a percentage) 9·8 ± 1·0%. Although the confidence intervals for a single cardiac output determination appear slightly smaller (±20%), there are fewer validation trials of the mitral method so there may be a publication bias in favour of encouraging results. For the pulmonary artery method, r values are significantly lower than for other methods: 0·80 ± 0·08, $P < 0·05$ compared with mitral and aortic studies. Similarly, SEE is greater: 1·44 ± 0·84 $l\,min^{-1}$ as is the SDD: 21·7 ± 4·9%, both $P < 0·01$ compared with the mitral and aortic methods. The 95% confidence intervals for a single estimate by the pulmonary artery method compared with standard methods would be ±2·9 $l\,min^{-1}$ or ±44%.

Only a few reports exist on tricuspid[10], oesophageal[31] and transtracheal[32] Doppler methods and it is too early to assess their reliability vs the aortic or mitral methods. Similarly the wide or dual beam pulsed wave method has yet to be adequately validated[33].

Thus, at rest in selected subjects, the aortic and, to a lesser extent, the mitral methods have been shown to be reasonably accurate compared with standard techniques, but quite wide confidence intervals have been reported for single measurements of cardiac output. The pulmonary artery method appears less reliable.

Blood velocity

To identify whether Doppler accurately measures blood velocity in humans as opposed to volume flow, as recorded by all the standards mentioned above, papers relating Doppler methods to electromagnetic catheters have been reviewed

Table 8 Validation studies of Doppler cardiac output measurement during exercise

First author	No.	Technique	Method	Timing	Range	r	slope	SEE or SDD	Analysis
Marx[65]	26 P	PW AA SS	RB	Sim.	6·0	0·86	0·93	1·4 (20%)	FFT Modal
Christie[25]	10	CW AA SS	TD	Sim.	4·8	0·78	0·71	3·8 (26%)	Max
Christie[25]	10	CW AA SS	Fick	Sim.	4·8	0·81	0·97	4·3 (30%)	Max
Ihlen[49]	20	PW AA SS	TD	Sim.				Supine 25W (16%)	Analogue
								Erect 25W (23%)	Analogue
								Erect 50W (22%)	Analogue
Maeda[62]	34	CW AA SS	TD	Sim.	3·0	0·84	1·33	1·22 (17%)	Max.

Abbreviations as for Table 5.

(Table 7). The correlation between the Doppler measurement and the electromagnetic catheter is good, and the SDD between the methods, 4–9% is less than for the cardiac output comparisons described above. Therefore, it appears that Doppler can measure intra-aortic blood velocity very accurately, but that there is increased uncertainty when this information is used to calculate volume flow.

Validation during exercise

Table 8 summarizes the validation papers during exercise. The reported correlations with other standards are significantly lower than comparisons at rest ($0{\cdot}82 \pm 0{\cdot}04$ compared with $0{\cdot}91 \pm 0{\cdot}06$, $P < 0{\cdot}01$), and the SDDs between methods are higher ($22{\cdot}0 \pm 4{\cdot}9\%$ compared with $13{\cdot}9 \pm 3{\cdot}4\%$, $P < 0{\cdot}001$), indicating greater disagreement between Doppler and standard methods during exercise, although whether the increased variation arises from the Doppler or the standard techniques is not clear.

All the quoted papers validate aortic Doppler methods, and little information exists for other recording sites. These are likely to be less accurate as the cross-sectional area as the mitral and tricuspid valve increases substantially during exercise and hence has to be measured at each recording, which has the associated difficulties of using echocardiography during exercise[25].

In summary, a single aortic Doppler estimate of cardiac output during exercise lies only within 95% confidence limits of ±44% compared with standard techniques. It does not appear that separate measurement of aortic root cross-sectional area for each measurement of cardiac output is essential because the area does not change appreciably during exercise[21,25].

Conclusions

The Doppler methods are reasonably accurate and fairly reproducible methods for measuring cardiac output non-invasively in selected subjects at rest, and there is little difference in accuracy between the aortic and mitral valve methods. The aortic method is simpler and less influenced by variations in cross-sectional area measurement, and has been more extensively validated during exercise.

References

[1] Hatle L, Angelsen B. Doppler ultrasound in cardiology. Philadelphia: Lea & Feibiger, 1985.

[2] Sahn DJ. Determination of cardiac output by echocardiographic Doppler methods: relative accuracy of various sites for measurement. J Am Coll Cardiol 1985; 6: 663–4.

[3] Nishimura RA, Miller FAJr, Callahan MJ, Benassi RC, Seward JB, Tajik AJ. Doppler echocardiography: theory, instrumentation, technique, and application. Mayo Clin Proc 1985; 60: 321–43.

[4] Pearlman AS. Evaluation of ventricular function using Doppler echocardiography. Am J Cardiol 1982; 49: 1324–30.

[5] Schuster AH, Nanda NC. Doppler echocardiographic measurement of cardiac output: comparison with a non-golden standard. Am J Cardiol 1984; 53: 257–9.

[6] Anon. Measurement of cardiac output. Lancet 1988; ii: 257–8.

[7] Light LH. Non-injurious ultrasonic technique for observing flow in the human aorta. Nature 1969; 224: 1119–21.

[8] Bojanowski LMR, Timmis AD, Najm YC, Gosling RG. Pulsed Doppler ultrasound compared with thermodilution for monitoring cardiac output responses to changing left ventricular function. Cardiovasc Res 1987; 21: 260–8.

[9] Huntsman LL, Gams E, Johnson CC, Fairbanks E. Transcutaneous determination of aortic blood flow velocities in man. Am Heart J 1975; 89: 605–12.

[10] Meijboom EJ, Horowitz S, Valdes-Cruz LM, Sahn DJ, Larson DF, Lima CA. A Doppler echocardiographic method for calculating volume flow across the tricuspid valve: correlative laboratory and clinical studies. Circulation 1985; 71: 551–6.

[11] Sequeira RF, Light LH, Cross G, Raftery EB. Transcutaneous aortovelography. A quantitative evaluation. Br Heart J 1976; 38: 443–50.

[12] Caro CG, Pedley TJ, Schroter RC, Seed WA. The mechanics of the circulation. Oxford: Oxford University Press, 1978.

[13] Seed WA, Wood NB. Velocity patterns in the aorta. Cardiovasc Res 1971; 5: 319–30.

[14] Schultz DL, Tunstall-Pedoe DS. Circulatory and respiratory mass transit. Ciba Symposium 1969; 172–202.

[15] Segadal L, Matre K. Blood velocity distribution in the human ascending aorta. Circulation 1987; 76: 90–100.

[16] Lewis JF, Kuo LC, Nelson JG, Limacher MC, Quinones MA. Pulsed Doppler echocardiographic determination of stroke volume and cardiac output: clinical validation of two new methods using the apical window. Circulation 1984; 70: 425–31.

[17] Sanders SP, Yeager S, Williams RG. Measurement of systemic and pulmonary blood flow and QP/QS ratio using Doppler and two-dimensional echocardiography. Am J Cardiol 1983; 51: 952–6.

[18] Goldberg SJ, Sahn DJ, Allen HD, Valdes-Cruz LM, Hoenecke H, Carnahan Y. Evaluation of pulmonary and systemic blood flow by 2-dimensional Doppler echocardiography using fast Fourier transform spectral analysis. Am J Cardiol 1982; 50: 1394–400.

[19] Touche T, de Zuttere D, Nitenberg A, Prasquier R, Gourgon R. Echo-Doppler quantitation of transmitral flow: a new method tested in adult patients. J Am Coll Cardiol 1985; 5: 425 (Abstr).

[20] Fisher DC, Sahn DJ, Friedman MJ *et al.* The mitral valve orifice method for noninvasive two-dimensional

echo Doppler determinations of cardiac output. Circulation 1983; 67: 872–7.
[21] Rassi AJr, Crawford MH, Richards KL, Miller JF. Differing mechanisms of exercise flow augmentation at the mitral and aortic valves. Circulation 1988; 77: 543–51.
[22] Stewart WJ, Jiang L, Mich R, Pandian N, Guerrero JL, Weyman AE. Variable effects of changes in flow rate through the aortic, pulmonary and mitral valves on valve area and flow velocity: impact on quantitative Doppler flow calculations. J Am Coll Cardiol 1985; 6: 653–62.
[23] Loeppky JA, Hoekenga DE, Greene ER, Luft UC. Comparison of noninvasive pulsed Doppler and Fick measurements of stroke volume in cardiac patients. Am Heart J 1984; 107: 339–46.
[24] Gardin JM, Tobis JM, Dabestani A *et al.* Superiority of two-dimensional measurement of aortic vessel diameter in Doppler echocardiographic estimates of left ventricular stroke volume. J Am Coll Cardiol 1985; 6: 66–74.
[25] Christie J, Sheldahl LM, Tristani FE, Sagar KB, Ptacin MJ, Wann S. Determination of stroke volume and cardiac output during exercise: comparison of two-dimensional and Doppler echocardiography, Fick oximetry, and thermodilution. Circulation 1987; 76: 539–47.
[26] Labovitz AJ, Buckingham TA, Habermehl K, Nelson J, Kennedy HL, Williams GA. The effects of sampling site on the two-dimensional echo-Doppler determination of cardiac output. Am Heart J 1985; 109: 327–32.
[27] Kitabatake A, Inoue M, Assao M *et al.* Noninvasive evaluation of the ratio of pulmonary to systemic flow in atrial septal defect by duplex Doppler echocardiography. Circulation 1984; 69: 73–9.
[28] Bland JM, Altman DG. Statistical methods for assessing agreement between two methods of clinical measurement. Lancet 1986; i: 307–10.
[29] Altman DG. Statistics and ethics in medical research: III How large a sample? Br Med J 1980; 281: 1336–8.
[30] Hills M, Armitage P. The two-period cross-over clinical trial. Br J Clin Pharmac 1979; 8: 7–20.
[31] Kumar A, Minagoe S, Thangathurai D *et al.* Noninvasive measurement of cardiac output during surgery using a new contiuous-wave Doppler esophageal probe. Am J Cardiol 1989; 64: 793–8.
[32] Abrams JH, Weber RE, Holmen KD. Continuous cardiac output determination using transtracheal Doppler: initial results in humans. Anesthesiology 1989; 71: 11–5.
[33] Looyenga DS, Liebson PR, Bone RC, Balk RA, Messer JV. Determination of cardiac output in critically ill patients by dual beam Doppler echocardiography. J Am Coll Cardiol 1989; 13: 340–7.
[34] Fraser CB, Light LH, Shinebourne EA, Buchthal A, Healy MJR, Beardshaw JA. Transcutaneous aortovelography: reproducibility in adults and children. Eur J Cardiol 1976; 4: 181–9.
[35] Gisvold SE, Brubakk AO. Measurement of instantaneous blood flow velocity in the human aorta using pulsed Doppler ultrasound. Cardiovasc Res 1982; 16: 26–33.
[36] Ihlen H, Endresen K, Myreng Y, Myhre E. Reproducibility of cardiac stroke volume estimated by Doppler echocardiography. Am J Cardiol 1987; 59: 975–8.
[37] Fagard R, Staessen J, Amery A. The use of Doppler echocardiography to assess the acute haemodynamic response to felodipine and metoprolol in hypertensive subjects. J Hypertension 1987; 5: 143–9.
[38] Mehta N, Boyle G, Bennett D *et al.* Hemodynamic response to treadmill exercise in normal volunteers: An assessment by Doppler ultrasonic measurement of ascending aortic blood velocity and acceleration. Am Heart J 1988; 116: 1298–307.
[39] Robson SC, Murray A, Peart I, Heads A, Hunter S. Reproducibility of cardiac output measurement by cross sectional and Doppler echocardiography. Br Heart J 1988; 59: 680–4.
[40] DeZuttere D, Touche T, Saumon G, Nitenberg A, Prasquier R. Doppler echocardiographic measurement of mitral flow volume: Validation of a new method in adult patients. J Am Coll Cardiol 1988; 11: 343–50.
[41] Robson SC, Boys RJ, Hunter S. Doppler echocardiographic estimation of cardiac output: analysis of temporal variability. Eur Heart J 1988; 9: 313–8.
[42] Hoit BD, Rashwan M, Watt C, Sahn DJ, Bhargava V. Calculating cardiac output from transmitral volume flow using Doppler and M-mode echocardiography. Am J Cardiol 1988; 62: 131–5.
[43] Chandraratna PA, Nanna M, McKay C *et al.* Determination of cardiac output by transcutaneous continuous-wave ultrasonic Doppler computer. Am J Cardiol 1984; 53: 234–7.
[44] Gardin JM, Burn CS, Childs WJ, Henry WL. Evaluation of blood flow velocity in the ascending aorta and main pulmonary artery of normal subjects by Doppler echocardiography. Am Heart J 1984; 107: 310–9.
[45] Nicolosi GL, Pungercic E, Cervesato E, Modena L, Zanuttini D. Analysis of interobserver and intraobserver variation of interpretation of the echocardiographic and Doppler flow determination of cardiac output by the mitral orifice method. Br Heart J 1986; 55: 446–8.
[46] McLennan FM, Haites NE, Mackenzie JD, Daniel MK, Rawles JM. Reproducibility of linear cardiac output measurement by Doppler ultrasound alone. Br Heart J 1986; 55: 25–31.
[47] Nicolosi GL, Pungercic E, Cervesato E *et al* Feasibility and variability of six methods for the echocardiographic and Doppler determination of cardiac output. Br Heart J 1988; 59: 299–303.
[48] Boyle G, Mehta N, Prindle K, Bennett ED. Ascending aortic blood velocity and acceleration measured by Doppler ultrasound in normal and beta-blocked subjects during treadmill exercise. Clin Sci 1986; 71: 150 (Abstr).
[49] Ihlen H, Endresen K, Golf S, Nitter-Hauge S. Cardiac stroke volume during exercise measured by Doppler echocardiography: comparison with the thermodilution technique and evaluation of reproducibility. Br Heart J 1987; 58: 455–9.
[50] Magnin PA, Stewart JA, Myers S, Von Ramm O, Kisslo JA. Combined Doppler and phased-array echocardiographic estimation of cardiac output. Circulation 1981; 63: 388–92.
[51] Alverson DC, Eldridge M, Dillon T, Yabek SM, Berman WJr. Noninvasive pulsed Doppler determination of cardiac output in neonates and children. J Pediatr 1982; 101: 46–50.

[52] Huntsman LL, Stewart DK, Barnes SR, Franklin SB, Colocousis JS, Hessel EA. Noninvasive Doppler determination of cardiac output in man. Clinical validation. Circulation 1983; 67: 593–602.

[53] Ihlen H, Amlie JP, Dale J *et al.* Determination of cardiac output by Doppler echocardiography. Br Heart J 1984; 51: 54–60.

[54] Nishimura RA, Callahan MJ, Schaff HV, Ilstrup DM, Miller RA, Tajik AJ. Noninvasive measurement of cardiac output by continuous-wave Doppler echocardiography: initial experience and review of the literature. Mayo Clin Proc 1984; 59: 484–9.

[55] Loeppky JA, Hoekenga DE, Greene ER, Luft UC. Comparison of noninvasive pulsed Doppler and Fick measurements of stroke volume in cardiac patients. Am Heart J 1984; 107: 339–46.

[56] Rose JS, Nanna M, Rahimtoola SH, Elkayam U, McKay C, Chandraratna PAN. Accuracy of determination of changes in cardiac output by transcutaneous continuous-wave Doppler computer. Am J Cardiol 1984; 54: 1099–101.

[57] Labovitz AJ, Buckingham TA, Habermehl K, Nelson J, Kennedy HL, Williams GA. The effects of sampling site on the two-dimensional echo-Doppler determination of cardiac output. Am Heart J 1985; 109: 327–32.

[58] Mehta N, Iyawe VI, Cummin ARC, Bayley S, Saunders KB, Bennett ED. Validation of a Doppler technique for beat-to-beat measurement of cardiac output. Clin Sci 1985; 69: 377–82.

[59] Vandenbogaerde JF, Scheldewaert RG, Rijckaert DL, Clement DL, Colardyn FA. Comparison between ultrasonic and thermodilution measurements in intensive care patients. Crit Care Med 1986; 14: 294–7.

[60] Innes JA, Mills CJ, Noble MIM *et al.* Validation of beat by beat pulsed Doppler measurements of ascending aortic blood velocity in man. Cardiovasc Res 1987; 21: 72–80.

[61] Morrow WR, Murphy DJJr, Fisher DJ, Huhta JC, Jeffereson LS, Smith EO. Continuous wave Doppler cardiac output: use in pediatric patients receiving inotropic support. Pediatr Cardiol 1988; 9: 131–6.

[62] Maeda M, Yokota M, Iwase M, Miyahara T, Hayashi H, Sotobata I. Accuracy of cardiac output measured by continuous wave Doppler echocardiography during dynamic exercise testing in the supine position in patients with coronary artery disease. J Am Coll Cardiol 1989; 13: 76–83.

[63] Zhang Y, Nitter-Hauge S, Ihlen H, Myhre E. Doppler echocardiographic measurement of cardiac output using the mitral orifice method. Br Heart J 1985; 53: 130–6.

[64] Coats AJS, Murphy C, Conway J, Sleight P. Validation of the beat to beat measurement of ascending aortic blood velocity in man using a new high temporal resolution Doppler ultrasound spectral analyser. Br Heart J 1990 (in press).

[65] Marx GR, Hicks RW, Allen HD, Kinzer SM. Measurement of cardiac output and exercise factor by pulsed Doppler echocardiography during supine bicycle ergometry in normal young adolescent boys. J Am Coll Cardiol 1987; 10: 430–4.

Non-invasive study of cardiac performance using Doppler ultrasound in patients with hypertension

P. J. LACOLLEY*, B. M. PANNIER*, B. I. LEVY** AND M. E. SAFAR*

*Department of Internal Medicine and *INSERM (U 337), Broussais Hospital and **INSERM (U 141), Lariboisière Hospital, Paris, France

KEY WORDS: Essential hypertension, Doppler, aortic acceleration, stroke distance.

Using a pulsed Doppler velocimeter with spectral analysis, it is possible to measure instantaneous ascending aortic blood velocity by the suprasternal approach. Cardiac output, stroke volume and maximal acceleration are evaluated from the aortic velocity curve. Maximal aortic acceleration is increased in patients with borderline hypertension by comparison with normal subjects and patients with sustained essential hypertension of the same age. Stroke distance is calculated as the ratio between stroke volume and the cross-sectional area of aortic valve measured by echocardiography. Stroke distance is significantly decreased in patients with sustained hypertension, suggesting that the distance covered by a column of blood passing through the aortic root during one cardiac cycle is smaller in patients with hypertension than in normal subjects.

Introduction

Echocardiographic and invasive haemodynamic studies using acute volume expansion have shown repeatedly that cardiac performance is normal or even increased in patients with essential hypertension[1–4]. However, these studies have been mostly global evaluations of cardiac performance, taking into account not only the heart but also the status of pre- and afterload. In this respect, the classical echocardiographic measures of left ventricular performance, such as ejection fraction and mean velocity of circumferential fibre shortening, are strongly influenced by modifications in afterload[5,6].

Doppler methods combined with echocardiography have been widely used for the evaluation of stroke volume and cardiac output in man[7–14]. From the study of aortic velocity curves, it has been possible to evaluate and validate maximal aortic acceleration[8,10–15] and to calculate stroke distance[8,11,12,16,17]. Both measurements have been suggested to be valid indices of cardiac contractility and performance[15,16,18–24]. Maximal aortic acceleration has been evaluated in spontaneously hypertensive rats and found to be normal or increased in younger animals and decreased in older rats, in association with an increase in cardiac mass[25]. In the present study, some applications are presented to evaluate cardiac output, stroke distance and maximal aortic acceleration in patients with borderline and sustained essential hypertension using non-invasive Doppler methods.

Address for correspondence: Professor Michel Safar, Department of Internal Medicine, Hopital Broussais, 96, rue Didot, 75674, Paris Cedex 14, France.

Basic concepts of the methodology

Ascending aortic blood velocity may be recorded with a hand-held transducer over the suprasternal notch, using a pulsed Doppler velocimeter as previously described[10,11]. Briefly, a pulsed range gated Doppler (ALVAR) transmits repeated bursts of ultrasound at a frequency of 4 MHz with a repetition rate of 10 kHz. After each burst, it is switched to receive reflected signals until the moment of the following ultrasound emission. An adjustable electronic gate is opened at a specific time after each burst of ultrasound and the reflected signals are recorded only during the time that the gate is opened. Signals are thus obtained at specific times which correspond to give distances from the transducer. Thus reflected signals are obtained from a 'gated' sample volume which can be moved to cover selected distances from the transducer. The sample volume can be located in the ascending aorta and positioned so as to exclude noise signals from walls and adjacent blood vessels, thus improving the accuracy of the measurement of mean blood velocity. A focused probe (diameter 5 mm) with a beam width of 3 cm diameter at a distance of 5 cm from the transducer may be used, since it illuminates most of the cross-sectional area of the aortic root. The duration of the gate, ranging from 1 to 3 s, is the smallest compatible with an acceptable signal to 'noise' ratio. The sample volume is thus a slice of about 3 cm diameter and 0·75–2·25 mm thickness. The sample volume is located at a distance ranging from 6 to 7·5 cm from the transducer. The frequency shift of reflected ultrasonic signals can be converted into velocity using the formula:

$$V = f^*c/2F\cos\theta \qquad [1]$$

where V is the velocity of the target in cm s^{-1}, f is the Doppler shift of frequency in Hz, θ is the angle between the ultrasound beam and the flow direction, F is the frequency of the emitted ultrasound and c is the speed of ultrasound in tissues in cm s^{-1}.

Single values for both the Doppler shift frequency and the angle are needed for the calculation of blood velocity. The mean Doppler shift frequency of the Doppler signal obtained from the ascending aorta is determined with spectral analysis using a fast Fourier transform analyser mounted in a HP Vectra PC. The averaged frequency is calculated for each 5 ms and converted to instantaneous blood velocity using the Doppler equation [1]. The transducer is pointed from the suprasternal notch toward the aortic valves. It is assumed that signal direction is close to axial and that θ is 18° (i.e. $\cos\theta$ is 0·95). Integrating velocity over one cardiac cycle gives a value proportional to the blood volume passing through the sample volume during the cardiac cycle.

The diameter of the sigmoid valve is measured by M-mode echocardiography. The product of the integrated velocity and the cross-sectional area of the aorta (CSA) gives the stroke volume (SV). Cardiac output (CO) can then be calculated as the product of the stroke volume and heart rate (HR). The integrated mean velocity over the time of the cardiac cycle is dimensionally a distance termed 'stroke distance' (SD). SD is related to the stroke volume by the equation

$$SD = \int_0^t V.dt = SV/CSA$$

Maximum acceleration (MA) defined as max dv/dt is calculated during the early period of ejection. Long-term (2 months) reproducibility of stroke distance and maximal acceleration were respectively $12{\cdot}0 \pm 2{\cdot}5$ and $14{\cdot}1 \pm 2{\cdot}1$%.

In previous studies[10,11], we have compared the Doppler method with the thermodilution and Fick techniques to determine cardiac output. The results indicated that measurements of left ventricular stroke volume and cardiac output by the Doppler method correlated well with thermodilution measurements over a wide range of cardiac output and stroke volume[10]. The equation for the regression line between cardiac output measurements by Doppler (DCO) and thermodilution (TDCO) was: $DCO = 0{\cdot}86\ TDCO + 0{\cdot}29\ l\ min^{-1}$ ($r = 0{\cdot}96$; $n = 26$).

Applications in essential hypertension

Non-invasive haemodynamic measurements were performed in normal subjects and in patients with borderline and sustained hypertension[26,27] for similar values for age, weight, height and body surface area. Table 1 summarizes Doppler measurements in normal (Group I) and hypertensive (Group II and III) subjects. Heart rate, left ventricular ejection time, stroke volume, cardiac output, cardiac index and stroke index did not differ in the three groups. Maximum aortic blood acceleration in normal subjects was 2200 ± 109 m s^{-2} and ranged between 1558 and 2867 ms^{-2}. Stroke distance in normal subjects was $14{\cdot}4 \pm 1{\cdot}0$ cm and ranged between 8·6 and 18·7 cm. Maximum aortic blood acceleration was similar in groups I and II and significantly higher in group III than in groups I ($P < 0{\cdot}01$) and II ($P < 0{\cdot}05$). Stroke distance was significantly reduced in group II subjects ($P < 0{\cdot}05$) but remained within the normal range in group III.

Several studies have demonstrated the reliability of the Doppler method as a non-invasive method to evaluate aortic blood flow velocity obtained from the suprasternal notch using pulsed Doppler and spectral analysis techniques[9,10,12]. However, to allow a more meaningful interpretation of the present non-invasive measurements, the intraobserver variability for stroke distance and maximal aortic acceleration was studied. Intraobserver variability after 3 months was 12% for stroke distance and 14% for maximal aortic acceleration (lower than the standard error of the mean). These results are in general agreement with those previously reported for comparable Doppler measurements[8,28]. The measurement of sigmoid aortic diameter by echocardiography is the most difficult to make but is necessary in order to obtain an accurate assessment of stroke volume or cardiac output[13,14,29,30]. An error in measurement of the diameter can cause a significant distortion in cardiac output values because the cross-sectional area is proportional to the square of the diameter. In this regard, it is important to note that stroke distance and maximum

Table 1 Non-invasive Doppler measurements in normal subjects (Group I) and patients with sustained (Group II) and borderline (Group III) essential hypertension. Mean age was respectively 37±2, 42±2 and 39±4 years

	Group I (n=11)	Group II (n=16)	Group III (n=11)
Mean arterial pressure (mmHg)	85±2·1	123±1·6	103±1·7
Heart rate (beats min^{-1})	62±1·8	69±4·0	65±2·2
Cardiac index (ml min^{-1} m^{-2})	3201±177	2877±239	3210±276
Stroke index (ml m^{-2})	54±4	42±4	49±3
Stroke distance (cm)	14·4±1·0	11·4±0·9*	12·0±0·06
Total peripheral resistance (mmHg.ml^{-1}.s^{-1})	0·89±0·08	1·41±0·09**	1·12±0·13
Maximum aortic acceleration (cm.s^{-2})	2200±109	2331±73	2656±100**

Data are mean ±1 standard error of the mean ***P*<0·01 vs Group I.

aortic acceleration are not influenced by diameter measurement and thus their reproducibility is only Doppler dependent.

On the basis of animal experiments, it has been suggested that maximum aortic blood flow acceleration is a sensitive index of left ventricular contractility which is relatively unaffected by preload and afterload[18,19]. In dogs, changes in left ventricular end-diastolic volume induced by different postural manoeuvres did not alter maximum acceleration[18]. In normal subjects, maximal aortic acceleration was found to be relatively insensitive to alterations of loading factors[15]. In the present study, the values of maximal aortic acceleration largely agreed with those of the literature[8–10,12,15,23], in particular with those involving comparison with the electromagnetic flowmeter[31,32]. Patients with borderline hypertension exhibited higher maximum aortic acceleration than normotensive controls. As shown experimentally[19], aortic acceleration is very sensitive to inotropic agents and may be enhanced by isoproterenol administration. On the other hand, increased sympathetic activity which predominantly affects beta-receptors is one of the most important features of patients with borderline hypertension[26,27]. Thus the higher maximum aortic acceleration in borderline hypertensive subjects might reflect a positive inotropic effect possibly related to an enhanced $beta_1$ sympathetic activity. In contrast, maximum aortic acceleration remained within the normal range in patients with sustained essential hypertension. The present findings are consistent with animal experiments[25] showing that young spontaneously hypertensive rats (SHR) with moderate left ventricular hypertrophy exhibited normal maximal acceleration of the aortic blood flow, whereas older SHR with higher cardiac mass have decreased maximal aortic acceleration.

The most important finding of the study was the decrease in stroke distance observed in patients with sustained essential hypertension. Haites *et al.*[16] have reported a decrease in stroke distance in hypertensive patients but nine of the fifteen studied patients were on treatment. Stroke distance is defined as stroke volume divided by aortic cross-sectional area (CSA) and is not influenced by body surface area[17]. Stroke distance has units of length (cm) and represents the distance covered by a column of blood through the aortic root during one cardiac cycle[16]. In the present study, this distance was smaller in patients with sustained essential hypertension than in normal subjects. This finding is difficult to interpret. The normal stroke volume in the presence of reduced stroke distance for a given cardiac cycle suggests an increase in aortic arch diameter, a finding which has been convincingly reported in hypertensive subjects[33,34]. However, decreased stroke distance might also imply a decrease in cardiac performance. Because stroke distance is unrelated to such morphologic characteristics of the subjects as body surface area (BSA) and aortic valve cross-sectional area (CSA), it may be considered a more sensitive index of cardiac performance than stroke index (depending on the CSA) and stroke volume (depending both on CSA and BSA). Decreased cardiac performance in the evolution of hypertension has been suggested by the animal studies of Pfeffer *et al.*[25], showing a progressive deterioration of cardiac function with age in SHR.

In conclusion, the present work suggested that cardiac performance evaluated from maximum aortic acceleration was increased in patients with borderline hypertension and remained within the normal range in patients with sustained essential hypertension. Another important finding was the significant decrease in stroke distance observed in patients with sustained essential hypertension. Whether this finding reflects a decreased cardiac performance is difficult to interpret and requires further investigation.

This study was performed with a grant from the Institut National de la Santé et de la Recherche Médicale (INSERM: U 337) and the Ministère de la Recherche. We thank Miss Brigitte Laloux for technical assistance and Mrs Annette Seban for presentation of the manuscript.

References

[1] Lutas EM, Devereux RB, Reis G *et al.* Increased cardiac performance in mild essential hypertension. (Left ventricular mechanics). Hypertension 1985; 7: 979–88.

[2] Leibson PR, Savage DD. Echocardiography in hypertension: a review I. Left ventricular wall mass, standardization, and ventricular function. Echocardiography 1986; 3: 181–218.

[3] Devereux RB, Savage DD, Sachs I, Laragh JH. Relation of hemodynamic load to left ventricular hypertrophy and performance in hypertension. Am J Cardiol 1983; 51: 171–6.

[4] Safar ME, London GM, Levenson JA, Simon ACh, Chau NP. Rapid dextran infusion in essential hypertension. Hypertension 1979; 1: 615–23.

[5] Mahler F, Ross J, O'Rourke RA, Covel JW. Effects of changes in preload, afterload and inotropic state on ejection and isovolumic phase measures of contractility in the conscious dog. Am J Cardiol 1975; 35: 626–34.

[6] Ross J. Cardiac function and myocardial contractility: a perspective. J Am Coll Cardiol 1983; 1: 52–62.

[7] Angelsen BAJ, Brubakk AO. Transcutaneous measurement of blood flow velocity in the human aorta. Cardiovasc Res 1976; 10: 368–79.

[8] Gisvold SE, Burbakk AO. Measurement of instantaneous blood flow velocity in the human aorta using pulsed Doppler ultrasound. Cardiovasc Res 1982; 16: 26–33.

[9] Goldberg SJ, Sahn DJ, Allen DH, Valdes-Cruz LM, Hoenecke H, Carnahan Y. Evaluation of pulmonary and systemic blood flow by 2-dimensional Doppler echocardiography using fast Fourier transform spectral analysis. Am J Cardiol 1982; 50: 1394–400.

[10] Levy BI, Tedgui A, Payen DM, Xhaard M, McIlroy MB. Non-invasive ultrasonic cardiac output measurement in intensive care unit. Ultrasound Med Biol 1985; 11: 841–9.

[11] Strauss A, Kedra W, Payen D, Levy B, Martineau DP. Non-invasive determination of stroke volume with pulsed Doppler. Echocardiography comparison with the Fick method. Herz 1986; 11: 269–76.

[12] Innes JA, Mills CJ, Noble MIM *et al.* Validation of beat by beat pulsed Doppler measurements of ascending aortic blood velocity in man. Cardiovasc Res 1987; 21: 72–80.

[13] Bouchard A, Blumlein S, Schiller NB *et al.* Measurement of left ventricular stroke volume using continuous wave Doppler echocardiography of the ascending aorta and M-mode echocardiography of the aortic valve. J Am Coll Cardiol 1987; 9: 75–83.

[14] Chandraratna PA, Nanna M, Mckay C *et al.* Determination of cardiac output by transcutaneous continuous-wave ultrasonic Doppler computer. Am J Cardiol 1984; 53: 234–7.

[15] Bennet ED, Barclay SA, Davis AL, Mannering D, Mehta N. Ascending aorta blood velocity and acceleration using Doppler ultrasound in the assessment of left ventricular function. Cardiovasc Res 1984; 18: 632–8.

[16] Haites NE, McLennan FM, Mowat DHR, Rawles JM. Assessment of cardiac output by the Doppler ultrasound technique alone. Br Heart J 1985; 53: 123–9.

[17] Mowat DHR, Haites NE, Rawles JM. Aortic blood velocity measurement in healthy adults using a simple ultrasound technique. Cardiovasc Res 1983; 17: 75–80.

[18] Noble MIM, Trenchard D, Guz A. Left ventricular ejection in conscious dogs: I. Measurement and significance of the maximum acceleration of blood from the left ventricle. Circ Res 1966; 19: 139–47.

[19] Nutter DO, Noble JO, Hurst VW. Peak aortic flow and acceleration as indices of ventricular performance in the dog. J Lab Clin Med 1971; 77: 307–18.

[20] Stein PD, Sabbah HN. Ventricular performance measuring during ejection. Studies in patients of the rate of change of ventricular power. Am Heart J 1976; 91: 599–606.

[21] Lambert CR, Nichols WW, Pepine CJ. Indices of ventricular contractile state: comparative sensitivity and specificity. Am Heart J 1983; 106: 136–44.

[22] Sabbah HN, Przybylski J, Albert DE, Stein PD. Peak aortic blood acceleration reflects the extent of left ventricular ischemic mass at risk. Am Heart J 1987; 113: 885–90.

[23] Sabbah HN, Gheorghiade M, Smith ST, Frank DM, Stein PD. Serial evaluation of left ventricular function in congestive heart failure by measurement of peak aortic blood acceleration. Am J Cardiol 1988; 61: 367–70.

[24] Gardin JM, Iseri LT, Elkayam U *et al.* Evaluation of dilated cardiomyopathy by pulsed Doppler echocardiography. Am Heart J 1983; 106: 1057–65.

[25] Pfeffer J, Pfeffer M, Fletcher P, Braunwald E. Alterations of cardiac performance in rats with established spontaneous hypertension. Am J Cardiol 1979; 44: 994–8.

[26] Safar M, Weiss Y, Levenson JA, London G, Milliez P. Hemodynamic study of 85 patients with borderline hypertension. Am J Cardiol 1973; 31: 315–9.

[27] Safar ME, Weiss YA. Borderline blood pressure elevation. In: Amery A, Ed. Hypertensive cardiovascular disease: Pathophysiology and treatment, Martinus Nijhoff, 1982: 365–76.

[28] Gardin JM, Dabestani A, Matin K, Allfie A, Russel D, Henry WL. Reproducibility of Doppler aortic blood flow measurements: studies on intraobserver,

inter-observer and day to day variability in normal subjects. Am J Cardiol 1984; 54: 1092–8.
[29] Loeppky JA, Hoekenga DE, Green ER, Luft UC. Comparison of non invasive pulsed Doppler and Fick measurement of stroke volume in cardiac patients. Am Heart J 1984; 107: 339–46.
[30] Ihlen H, Amlie J, Dale J *et al.* Determination of cardiac output by Doppler echocardiography. Br Heart J 1984; 51: 54–60.
[31] Mehta N, Noble M, Mills C, Pugh S, Drake-Holland A, Bennet D. Doppler measured ascending aortic blood velocity and acceleration. Validation against an electromagnetic catheter-tip system in humans. Circulation 1987; 76: IV-94 (Abstr).
[32] Stein PD, Sabbah HN, Albert DE, Snyder JE. Blood velocity and acceleration: comparison of continuous wave Doppler with electromagnetic flowmetry. Fed Proc 1985; 44: 1565 (Abstr).
[33] Merillon JP, Fontenier GJ, Lerallut JF *et al.* Aortic input impedance in normal man and arterial hypertension: its modification during changes in aortic pressure. Cardiovasc Res 1982; 16: 646–56.
[34] Isnard RN, Pannier BM, Laurent ST, London GM, Diebold B, Safar ME. Pulsatile diameter and elastic modulus of the aortic arch in essential hypertension: a noninvasive study. J Am Coll Cardiol 1989; 13: 399–405.

Measurement of cardiac output by M-mode and two-dimensional echocardiography: application to patients with hypertension

D. C. WALLERSON, A. GANAU, M. J. ROMAN AND R. B. DEVEREUX

Department of Medicine, The New York Hospital-Cornell Medical Center, New York, New York and the Institute of Clinical Medicine, University of Sassari, Sassari, Italy

KEY WORDS: Echocardiography, hypertension, Doppler, ventricular wall thickness, Goldblatt hypertension, rats.

The development of echocardiography has permitted non-invasive estimation of left ventricular stroke volume and cardiac output, albeit with some limitations. These approaches are particularly suited to hypertensive patients.

Introduction

Echocardiography has become established as a relatively inexpensive and reliable non-invasive cardiac diagnostic modality. These properties make it an attractive tool to screen large populations for suspected cardiac disease[1]. Echocardiography has been shown to be the most sensitive method for clinical detection of left ventricular (LV) hypertrophy[2,3], making it particularly suitable for evaluation of hypertensive patients[4]. Recent research indicates that increased echocardiographic LV mass is the strongest predictor, other than age, of increased cardiac morbidity and mortality in hypertensive patients[5,6] and members of the general population[7,8]. These observations suggest that echocardiography will play an increasingly important role in risk-stratification, guiding some therapeutic decisions in hypertensive patients.

The results of other echocardiographically based studies suggest that non-invasive measurements of haemodynamic variables are also related to the pathophysiology of systemic hypertension. Thus, echocardiographically derived estimates of peripheral resistance (but not measurements of arterial pressure) were found to add to the ability of LV mass to predict complications of hypertension[9]. Furthermore, clinical and experimental studies indicate that the level of volume load, as measured by stroke volume or cardiac output, contributes as strongly as blood pressure to the determination of LV mass[10,11]. Therefore, it is desirable to include reliable measurements of cardiac output in investigations of the pathophysiology of hypertension and its effects on the cardiovascular system. Because of the widespread clinical and research use of echocardiography to assess LV geometry in hypertension, it would be convenient if haemodynamic measurements could be made by the same technique.

An ideal method to measure cardiac output should be inexpensive, non-invasive, give accurate and reproducible measurements, and be well tolerated for serial determinations. Echocardiography satisfies the requirements for reasonable cost, non-invasiveness and serial applicability. Accordingly, this review primarily examines the accuracy and reproducibility of M-mode and two-dimensional (2-D) echocardiographic methods of measuring stroke volume and cardiac output. Because most contemporary echocardiographs have Doppler capability, data on Doppler assessment of cardiac output will be presented briefly for comparison.

Measurement of cardiac output by M-mode echocardiography

M-mode echocardiography is the oldest and simplest ultrasonic method to measure cardiac output. The primary measurements utilized are LV internal dimensions measured between the papillary muscle and mitral leaflet tips. Although several approaches to timing of measurements and recognition of interfaces have been used, by the widely

Supported in part by grant HL-18323 from the National Heart, Lung and Blood Institute, Bethesda, Maryland.

Address for correspondence: Donald C. Wallerson, M.D., Division of Cardiology, Box 222, The New York Hospital-Cornell Medical Center, 525 East 68th Street, New York, N.Y. 10021, U.S.A.

Table 1 Results of validation studies of M-mode echocardiographic measurements of stroke volume

Study	N	Reference method	r	SEE	Mean difference	Echo method	Subject characteristics
Popp[13]	30	Fick	0·97	7 ml	−1 ml	Cube	Varied heart disease without valvular regurgitation
Pombo[14]	27	Biplane angiography	0·83	25 ml	+8 ml	Cube	Valvular disease, CAD, cardiomyopathies
ten Cate[15]	35	Single plane angiography	0·82	20 ml	−5 ml	Cube	21/35 had CAD or valvular heart disease
Kronik[16]	25	Contrast angiography	0·74	18 ml	—	Teichholz	CAD, Congestive COM, some without cardiac disease
Kronik[16]	36	Thermodilution	0·86	12 ml	—	Teichholz	CAD, Congestive COM, some without cardiac disease
Feigenbaum[17]	42	Biplane angiography	0·72	32 ml	−17 ml	Cube	Not described
Gibson[18]	50	Biplane angiography	0·91	61 ml	—	Cube	Valvular disease, IHSS, cardiomyopathies, ventricular aneurysms
Belenkie[19]	18	Single plane angiography	0·44	55 ml	+6 ml	Cube	Valvular disease, five with cardiomyopathies
Ludbrook[20]	15	Biplane angiography	0·70		+29 ml	Cube	Patients without asynergy
Teichholz[21]	12	Single plane angiography	0·97			Teichholz	Patients without asynergy
Teichholz[21]	12	Single plane angiography	0·96			Cube	Patients without asynergy
Teichholz[21]	12	Single plane angiography	0·25			Cube	Patients with asynergy
Kiowski[22]	10	Dye dilution	0·72		−2 ml	Cube	Normals
Present Study	39	Doppler	0·62	13 ml	−3 ml	Teichholz	Normal or hypertensive
Present Study	100	2D echo stroke volume	0·72	12 ml	+8 ml	Teichholz	Normal or hypertensive

Abbreviations: CAD = coronary artery disease; COM = cardiomyopathy; IHSS = idiopathic hypertrophic subaortic stenosis; SEE = standard error of the estimate.

used American Society of Echocardiography recommendations[12], LV end-diastolic diameter is measured at the onset of the QRS complex and the end-systolic dimension is measured at the nadir of posterior septal motion by the leading edge to leading edge technique. Early investigators recognized that these linear dimensions could be cubed to provide reasonable estimates of the volume of normally-shaped left ventricles — based on the fact that such ventricles resemble a prolate ellipsoid with a long-axis twice the length of the short axes[13,14]. LV volumes derived by this method were shown to parallel closely those obtained by contrast angiography[13–15], allowing derivation of reasonable estimates of stroke volume in patients with valvular disease or other conditions associated with normal LV geometry[16–22], with less good results in patients with ventricular asynergy[18,21] (Table 1).

This method is simple, but has a number of limitations. First, since the ventricle becomes more spherical as it dilates, the ratio of LV long-axis dimension to short-axis dimension decreases; conversely, when the LV is underfilled, it becomes more exaggeratedly elliptical[23]. As a result, the simple cube function method tends to overestimate the volume of enlarged ventricles and to underestimate that of small chambers. The most widely used method to correct for this tendency was developed by Teichholz *et al.*[21] in which

$$\text{Volume} = \left(\frac{7{\cdot}0}{2{\cdot}4 + D}\right) \times D^3 \qquad [1]$$

where D = left ventricular internal diameter (LVID) in cm.

Second, this method assumes that global LV function is accurately reflected by contraction of the limited segment sampled for measurement of LV minor-axis dimensions. Reasonably accurate measurement of stroke volume and cardiac output by this method has been obtained in subjects with symmetrically contracting ventricles. The standard error of the estimate of stroke volume in such patients has been reasonably low compared with several reference standards, including thermodilution or contrast angiographic measurements (12

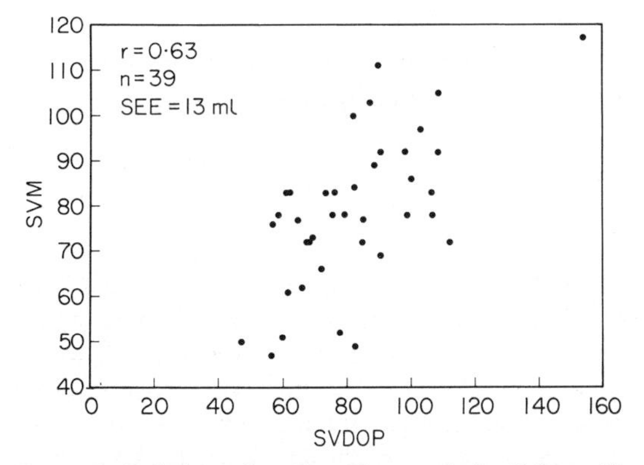

Figure 1 Relation of stroke volume calculated from M-mode echocardiographic measurements by the formula of Teichholz *et al.*[21] (vertical axis) and from Doppler echocardiographic measurements by the method of Dubin *et al.*[24] (horizontal axis) in 27 normotensive and 12 hypertensive adults.

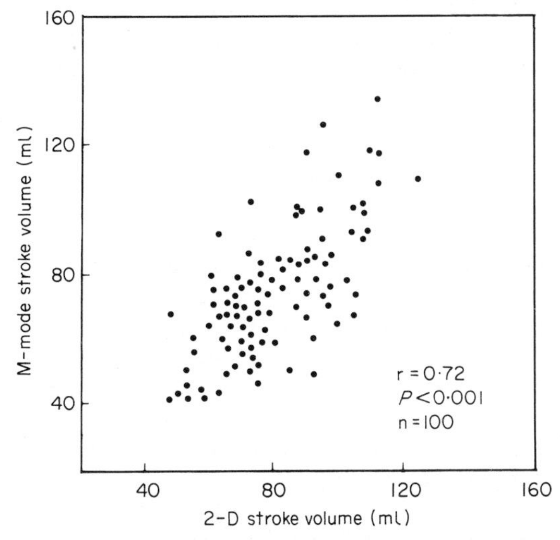

Figure 2 Relation of stroke volume derived by M-mode echocardiography (vertical axis) and by two-dimensional echocardiography using the biplane ellipsoidal formula[10] (horizontal axis) in 50 normotensive and 50 mildly hypertensive adults.

and 18 ml, respectively[16]) or non-invasive Doppler or 2-D echocardiographic measurements in our laboratories (13 and 12 ml, respectively) (Figs 1 and 2). However, in many patients with coronary artery disease, congenital heart disease or pulmonary hypertension segmental wall motion abnormalities may exist that render M-mode measurements of the length and shortening of a single axis unrepresentative of overall LV size or function. Several reports showing low accuracy of M-mode echocardiographic stroke volume measurements in patients with previous myocardial infarction confirm this limitation[17,21,25]. A third, the conceptual limitation of extrapolating from a single linear dimension to three-dimensional measurements of stroke volume and cardiac output, is that any error in primary measurements will be cubed in the calculation.

The data summarized in Table 1 permit one to assess the usefulness of M-mode echocardiographic measurements of LV stroke volume in patients with symmetrically contracting left ventricles. With standard errors of the estimate of 12 to 18 ml compared with various invasive and non-invasive reference standards (none of which is completely error-free), 95% confidence intervals for detection of differences between patients would be ± 24 to 35 ml. This indicates that moderate to large differences between individuals in stroke volume can be detected by this method, but that small ones cannot. Approximate 95% confidence intervals for mean stroke volumes in groups would be 8 to 12 ml for N = 10, 4 to 6 ml for N = 30 and 2 to 4 ml for N = 100. These data indicate that M-mode echocardiography measures stroke volume with moderate accuracy in patients with symmetric LV contraction, allowing detection of relatively small differences between or over time within moderate to large groups of patients. However, this technique has too great a variability to detect small differences between or within subjects.

Measurement of left ventricular volumes by two-dimensional echocardiography

When 2-D echocardiography was initially developed, it was expected that the advantages of more complete spatial information would improve the accuracy of measurements of cardiac chamber volumes. Variations of chamber shape could be clearly appreciated and more complex and appropriate geometric models of the left ventricle would permit more accurate volume measurements. The specific advantages of 2-D imaging included: (1) the ability to measure the LV long-axis directly; (2) the ability to orient short-axis imaging planes correctly vis-a-vis the long-axis and to ensure the choice of a correct level in the ventricle (e.g., papillary muscle tips); and (3) by making area rather than linear dimension the primary measurement, results would become less sensitive to errors due to discrepancies between the irregular shape of many ventricles and the idealized configuration implied in simple geometric models.

Models of LV shape have been combined with primary 2-D echocardiographic measurements to estimate LV volumes, the necessary prerequisite to calculation of stroke volume. As may be seen in Table 2, many models have been compared with a variety of in vivo and in vitro reference standards[26–31]. Most studies validating 2-D echocardiographic volume measurements are experimental comparisons between echocardiographic and invasively determined end-systolic and end-diastolic volumes under highly controlled (and artificial) circumstances. Extrapolation of the promising results of most of these studies to the clinical situation may overestimate the accuracy of the methods.

METHODS BASED ON SIMPLE GEOMETRIC MODELS

Methods based on prolate ellipsoidal geometry have proven the most popular (Fig. 3). The long-axis dimension (LAX) is usually measured from the apical four-chamber view, or less frequently the apical two-chamber view. The minor axes (D1 and D2) may be directly measured or a common minor axis may be derived as the mean radius from measured cavity area in the short-axis parasternal view; less frequently the minor axis is measured from apical views. Measurements are taken at the black-white interface of the endocardial border. The equation for volume (V) of a length-diameter prolate ellipsoid is:

$$V = 4/3\pi\left(\frac{L}{2}\right)\left(\frac{D1}{2}\right)\left(\frac{D2}{2}\right) \qquad [2]$$

If the minor axis dimensions are assumed equal this formula may be simplified to:

$$V = \frac{\pi}{6} L \times D^2 \qquad [3]$$

A slight modification of this method is the prolate ellipsoid length-area method. The inherent irregularities of LV shape should have less effect on measurements if ventricular areas rather than radii are directly measured. Multiple investigations have evaluated this geometric model[26], which utilizes two variations:

$$(1) \quad V = \frac{\pi}{6}(L)\left(\frac{4A4}{\pi L}\right)\left(\frac{4A1}{\pi D1}\right) \qquad [4]$$

where L and A4 are length and area directly measured from apical four-chamber views and D1 and A1 are diameter and area directly measured from parasternal short axis views at the level of the mitral valve tips (Fig. 3). Alternatively,

$$(2) \quad V = \frac{\pi}{6}(L)\left(\frac{4A4}{\pi L}\right)\left(\frac{4A2}{\pi L2}\right) \qquad [5]$$

where A4 and L are as above and A2 and L2 are measured from the orthogonal two-chamber apical views (Fig. 3). Conceptually, equation 5 would seem preferable since the areas are measured from presumably orthogonal planes that should include the long-axis plane of the left ventricle. An internal check of the accuracy of measurements used by this method is provided by the fact that L should be equal to L2. This is particularly important because the true LV apex is commonly missed in the four-chamber view[32], resulting in underestimation of chamber length and area, and as a consequence underestimation of LV volumes.

LV volume can also be calculated by a single-plane area-length prolate ellipsoid model, using the equation:

$$V = \frac{8\,(A4)^2}{3\,\pi L} \qquad [6]$$

where A4 and L are again as defined above (Fig. 3). Studies validating this method of LV volume measurement are shown in Table 2.

As expected conceptually, the biplane volume methods seem more accurate than the single-plane methods and the biplane methods with the largest number of primary measurements are the most accurate (Table 2).

METHODS BASED ON SIMPSON'S RULE

Simply stated, Simpson's rule proposes that the volume of a large figure may be calculated as the sum of a series of smaller figures which when placed together reproduce the larger figure. Fig. 4 shows how this rule is applied to the left ventricle by dividing the chamber along its long-axis into a series of roughly cylindrical slices. In vitro studies of perfused hearts show that this method gives extremely accurate measurements of LV volume when a large number of 'slices' are imaged, with only modest loss of accuracy with simplification to use only five parallel tomographic views[27]. However, direct application of Simpson's rule is not possible in adults, since interposed ribs and lungs limit available acoustic windows and prevent imaging in evenly

Table 2 Result of validation studies of two-dimensional echocardiographic measurements of left ventricular volumes

Author	N	Reference standard	Echocardiographic geometric model	Volume compared	r	SEE	Subject characteristics
Gehrke[26]	20	Single plane angiography	Area length (P-E)	EDV and ESV	0·88	57 ml	Valvular disease, HCM, and dilated COM
Eaton[27]	6	Direct volume measurement	Simpson's Rule	EDV and ESV	0·97	2·9 ml	Six dog hearts suspended in blood solution; range 9·5–54·7 ml
Schiller[28]	30	Biplane angiography	Modified Simpson's Rule	EDV	0·80		16/30 with wall motion abnormalities
				ESV	0·90		
Folland[29]	35	Biplane angiography	Modified Simpson's Rule	EDV	0·76	43 ml	26 with CAD; four with valvular disease
				ESV	0·86	32 ml	
			Biplane	EDV	0·67	49 ml	26 with CAD; four with valvular disease
				ESV	0·72	44 ml	
			Ellipsoid single plane	EDV	0·61	52 ml	26 with CAD; four with valvular disease
				ESV	0·64	48 ml	
			Biplane hemi-cylinder	EDV	0·68	49 ml	26 with CAD; four with valvular disease.
				ESV	0·75	42 ml	
Wyatt[30]	21	Direct volume measurement	Simpson's Rule	EDV and ESV	0·98	6·6 ml	Formalin-fixed dog hearts
			Area length hemi-cylinder/cone	EDV and ESV	0·97	10·9 ml	
			prolate ellipsoid	EDV and ESV	0·97	8·6 ml	
Silverman[31]		Biplane angiography	Simpson's Rule	EDV	0·95	12 ml	20 children (2 months to 18 years)
			Simpson's Rule	ESV	0·89	7 ml	
			Four-chamber area length (P-E)	EDV	0·95	12 ml	Mixed congenital and valvular diseases
			Four-chamber area length (P-E)	ESV	0·92	6 ml	
			Biplane area length (P-E)	EDV	0·94	15 ml	Mixed congenital and valvular diseases
			Biplane area length (P-E)	ESV	0·91	7 ml	
			Two-chamber area length (P-E)	EDV	0·92	16 ml	Mixed congenital and valvular diseases
			Two-chamber area length (P-E)	ESV	0·89	8 ml	

Abbreviations: COM = cardiomyopathy; EDV = end-diastolic volume; ESV = end-systolic volume; HCM = hypertrophic cardiomyopathy; P-E = prolate ellipsoid; SEE = standard error of the estimate.

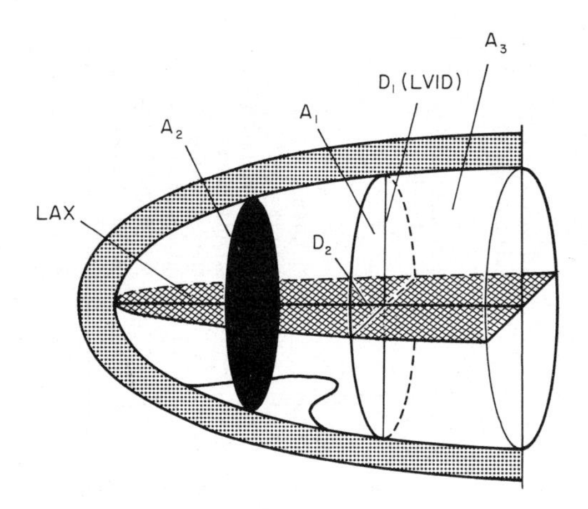

Figure 3 The prolate ellipsoid model of the left ventricle. The long-axis dimension (LAX), the two mid-ventricular short-axis dimensions (D1 and D2), long-axis (cross-hatched area, A4) and short-axes (A1 and A2) are illustrated.

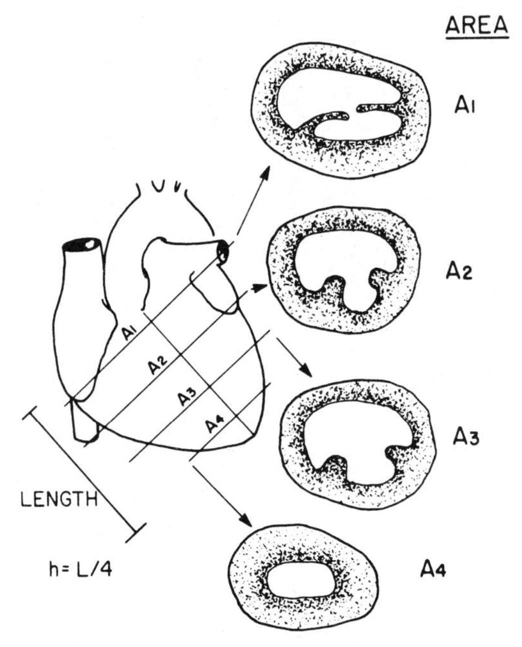

Figure 4 Diagram illustrates the division of the left ventricle into a series of cylinders using a modified Simpson's Rule approach. The volume of the basal quarters of the ventricle of the apex is determined as the volume of an ellipsoid volume segment. (From Wyatt *et al.*[33] Reproduced by permission of the American Heart Association).

spaced parallel planes, hence the modification described in Fig. 4.

There is, however, an adaptation of Simpson's Rule that is currently feasible. The apical two- and four-chamber views are used to image the left ventricle in approximately orthogonal planes that should share a common long-axis. The chamber is then divided into a series of even slices arranged perpendicular to the long-axis. Each slice is thus an ellipsoid cylinder of a height (H) determined by dividing the long-axis dimension by the number of slices. The two orthogonal short-axis diameters at any point along the LV long-axis may be directly measured from the two apical views, again using the black-white interface method. The volume of the entire chamber is calculated as:

$$V = \frac{\pi}{4} H \sum_{0}^{N} D1 \times D2 \qquad [7]$$

This method is especially attractive since it does not require that the LV shape conform to any specific geometric model. It is therefore flexible in measuring the volume of chambers with distorted or irregular shape, provided that the number of slices is large enough to describe these irregularities faithfully. Furthermore, the advent of readily available sophisticated computer software has automated the necessary computations. Investigations using this method have shown excellent correlations with low standard errors of estimate of chamber volumes (3 to 7 m) compared with isolated, perfused beating hearts or formalin-fixed hearts (Table 2).

ESTIMATION OF LEFT VENTRICULAR VOLUME BY COMBINED GEOMETRIC MODELS

The final geometric models to estimate LV volume employ short-axis imaging planes to divide the ventricle into segments of different shape. The volumes of each segment are measured separately and then summed to provide the total LV volume. The geometric figures most commonly used for components are the cylinder, the cone, the truncated cone and the ellipsoid segment (Fig. 5). Fig. 5e, f and g illustrate, respectively, the cylinder-hemiellipse, the cylinder-truncated cone-cone and the cylinder-cone geometric models. In each of these models, the long-axis length is measured from an apical two- or four-chamber view and the short-axis dimensions are measured at the level of the mitral valve (A1) and at the papillary muscle (A2). It is assumed that these two planes divide the LV long-axis into three equal segments so that 'H', the height of each segment, is equal to one third of

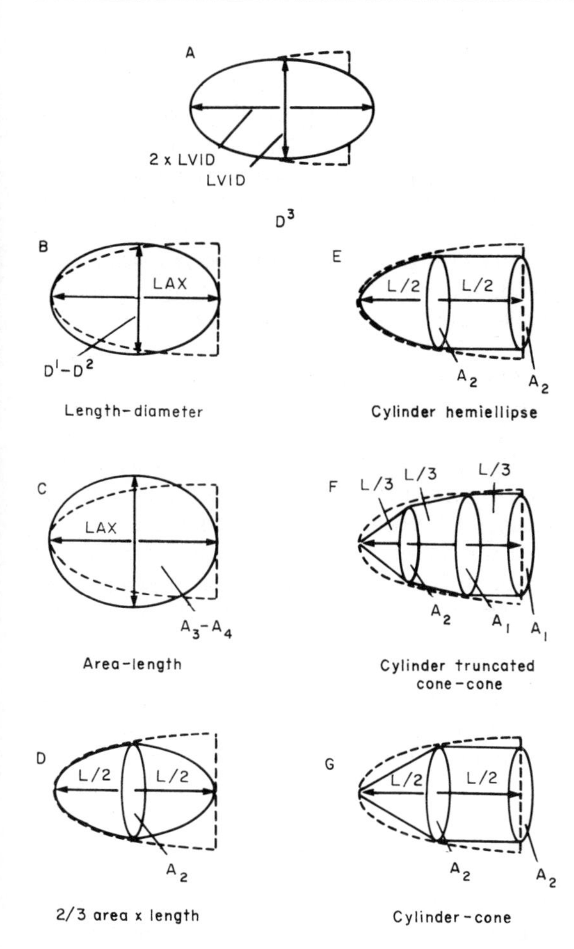

Figure 5 Series of diagrams illustrating the primary geometric models employed in the two-dimensional measurement of left ventricular volumes.

the long-axis. The simplest of these methods, the cylinder-cone (Fig. 5g) has already been discussed as a modification of the prolate ellipsoid (Fig. 5d).

The cylinder-truncated cone-cone model (Fig. 5f) is represented by the volume equation:

$$V = \frac{L}{3}(A1) + \frac{I}{3}\left(\frac{(A1+A2)}{2}\right) + \frac{I}{3}(A2)\frac{L}{3} \qquad [8]$$

Volumes calculated using this model have correlated well with single plane angiographic volumes in clinical studies (Table 2).

In an in vitro study, LV chamber volume calculated from 2-D echocardiographic measurements and the cylinder-hemi-ellipsoid model (Fig. 5e) compared favourably to directly measured volumes with very small standard errors of estimate[30] Unfortunately, a clinical study by Folland *et al.*[29] showed a less satisfactory correlation with biplane angiography. The formula for LV volume from this model is:

$$V = 5/6A2 \times L \qquad [9]$$

This formula has worked well enough for it to remain a viable option for measurement of chamber volumes by 2-D echocardiography. An adaptation of this formula has been demonstrated to give highly accurate measurements of LV muscle mass in both clinical[34] and experimental[35] studies.

Two-dimensional echocardiographic measurements of stroke volume

Although 2-D echocardiographic volume measurements have been used extensively to measure LV ejection fraction, there has been little use of this technique to assess stroke volume and cardiac output. Data are available, however, concerning normal values for stroke volume by several 2-D echocardiographic methods in adults and children[31,36–39] (Table 3) and on the relation between stroke volume determined by one of these methods and LV mass (Fig. 5).

NORMAL VALUES OF TWO-DIMENSIONAL ECHOCARDIOGRAPHIC STROKE VOLUME

Several groups of investigators have reported normal values for LV volume or stroke volume measured by 2-D echocardiography in volunteers who were free of clinically apparent heart disease. Unfortunately, in many studies primary data are presented only for measurements of end-diastolic and end-systolic LV volume, permitting only calculation of mean stroke volume for the group. Furthermore, full characterization of the demographic characteristics, body build or blood pressure of these subjects is rarely available.

The data summarized in Table 3 show that LV stroke volume in normal subjects tends to be relatively similar despite use of a variety of 2-D methods, with a value for mean stroke index between 32 and 40 ml m^{-2} except for the 47 ml m^{-2} obtained by Wahr *et al.* in males by one 2-D method[39] (Table 3). For most 2-D echocardiographic methods, the reported values of stroke index have been at or below the bottom of the range of mean normal values obtained with invasive methods (40–45 ml m^{-2}). This may in part reflect errors in measurement or limitations of the chosen geometric models. However, this may also reflect a more fundamental propensity of 2-D echocardiography to underestimate chamber area or volume.

Two lines of evidence support this hypothesis. First, a meticulous in vitro study by Helak *et al.*[40] demonstrated that directly measured LV cavity area was systematically underestimated by 2-D echocardiography, with differences among echographs in the magnitude of this error. A more recent study by Pini *et al.*[41] showed that the underestimation of cavity area was reduced, but not eliminated, with the use of newer equipment, with the best results obtained using an annular array transducer. An even more recent study from our laboratory (Mensah FA, Pini R *et al.*, unpublished data) showed that the cavity volume of in vitro objects was accurately predicted ($r = 0{\cdot}995$) but also systematically underestimated (by a mean of 3·5%) in three-dimensional volume images, constructed from primary 2-D recordings by a computer-assisted method.

A second reason for 2-D echocardiographic underestimation of LV chamber volume is that the chamber's long-axis is often not fully visualized[32], resulting in reduced measurements of chamber volume. The chances of successfully obtaining the true long-axis dimension may be improved by several manoeuvers, including locating the point of maximal impulse and then moving the transducer laterally and inferiorly to maximize the long-axis dimension. When this maximum is obtained in the four-chamber view, a 90° rotation of the transducer should be performed to see if the orthogonal two-chamber view shares the maximized long-axis dimension. If it does not (i.e. two-chamber long-axis dimension is larger), it implies that the original long-axis dimension was understated. In addition, recent studies using three-dimensional imaging systems have shown that conventional 2-D echocardiographic imaging planes are commonly misoriented by 15–20°, or even more, both with regard to the true LV long- and short-axes[42,43] and with regard to divergences from the assumed 90° difference between supposedly orthogonal pairs of views[44].

REPRODUCIBILITY OF 2-D ECHOCARDIOGRAPHIC VOLUME MEASUREMENTS

Of all measurements obtained by 2-D echocardiography, volumes show the greatest variability. Gordon *et al.*[37] found 95% confidence limits to be ±15% for end-diastolic volume and ±25% for end-systolic volume. Himelman *et al.*[45] similarly showed 95% confidence limits of 11% and 15% for the end-diastolic and end-systolic volumes; the 95% confidence limits for stroke volume by 2-D echocardiography was similar at 14%. Group data from the study by Gordon *et al.* showed that mean changes of 2% for end-diastolic volume and 5% for end-systolic volume would be significant in a population of 30 subjects. Once again, similar analyses are not available for stroke volume.

RELATION OF TWO-DIMENSIONAL ECHOCARDIOGRAPHIC STROKE VOLUME TO LEFT VENTRICULAR MASS

LV hypertrophy, defined as an increase in LV mass, is a common response of the heart to hypertension. LV mass may be calculated accurately from M-mode echocardiographic measurements of end-diastolic LV wall thicknesses and chamber diameter by the simple cube function formula, using necropsy-validated regression equations[2,46,47]. Contrary to conventional assumptions, hypertensive cardiac hypertrophy may be either concentric, where wall thicknesses are increased disproportionately to LV internal dimension, or eccentric, where the wall thickness/chamber radius ratio remains within the normal range associated with an increase to high-normal or frankly elevated values of LV internal dimension. Many clinical and experimental investigations have shown that an increased cardiac output is an important component of the haemodynamic load presented by both borderline[48–50] and sustained essential hypertension[51,52]. A recent study from our laboratory of rats with Goldblatt hypertension showed that those with the one-kidney one-clip form, a model known to be associated with volume expansion, had greater cardiac output and higher LV mass than either animals with renin-dependent 2-kidney one-clip hypertension or control rats[11]. In this experiment, stroke volume was measured by M-mode echocardiography. In a clinical study, Ganau *et al.*[10] found that LV mass was positively related to stroke index and, in a multivariate analysis, stroke index, systolic blood pressure and end-systolic stress/volume index ratio were each independently related to left ventricular mass and accounted for 66% of its overall variability. This study used 2-D echocardiography to measure stroke volume by the length-diameter prolate ellipsoid formula given above (equation 2). The relation between 2-D echocardiographic stroke volume by this method and LV mass is illustrated in Fig. 6.

Table 3 Two-dimensional echocardiographic measurement of stroke volume validation studies

Study	N	Reference standard	Geometric model	Stroke index	r vs reference standard	SEE	Mean difference	Subject characteristics
Erbel[36]	35	None	Area-length prolate ellipsoid	40 ± 7 ml m^{-2}	—	—	—	Normal men
Erbel[36]	20	None	Area-length prolate ellipsoid	36 ± 11 ml m^{-2}	—	—	—	Normal women
Gordon[37]	20	None	Prolate ellipsoid 8A2/3 L	32 ml m^{-2}	—	—	—	Normal men and women
Gordon[37]	10	None	Area-length prolate ellipsoid	33 ml m^{-2}	—	—	—	CAD
Byrd[38]	44	None	Modified Simpson's Rule	40 ml m^{-2}	—	—	—	44 men
Byrd[38]	30	None	Modified Simpson's Rule	34 ml m^{-2}	—	—	—	30 women
Wahr[39]	29	None	Two-chamber area-length P-E	47 ml m^{-2}	—	—	—	Normal male volunteers
			Four-chamber area-length P-E	40 ml m^{-2}	—	—	—	Normal male volunteers
			Biplane two-chamber + four-chamber P-E	44 ml m^{-2}	—	—	—	Normal male volunteers
			Biplane two-chamber + short-axis P-E	43 ml m^{-2}	—	—	—	Normal male volunteers
			Modified Simpson's Rule	40 ml m^{-2}	—	—	—	Normal male volunteers
Wahr[39]	23	None	Two-chamber area-length P-E	18 ml m^{-2}	—	—	—	Normal female volunteers
			Four-chamber area length P-E	36 ml m^{-2}	—	—	—	
			Biplane two-chamber + four-chamber P-E	38 ml m^{-2}	—	—	—	
			Biplane two-chamber + short axis P-E	37 ml m^{-2}	—	—	—	
			Modified Simpson's Rule	32 ml m^{-2}	—	—	—	
Silverman[31]	20	Biplane angiography	Simpson's Rule		0·92	7 ml	−8·1 ml	Mixed congenital and valvular disease; 20 children (2 months to 18 years)
			Four-chamber area-length P-E		0·94	9 ml	−6·4 ml	
			Biplane area-length P-E		0·95	10 ml	−5·2 ml	
			Two-chamber area-length P-E		0·93	10 ml	−7·3 ml	
Present study	100	M-mode echo (Teichholz)	Length-diameter prolate ellipsoid	72 ± 20 ml	0·72	12 ml	−8 ml	50 normals, 50 hypertensives; both sexes
Present study	50	None		38 ± 8 ml m^{-2}	—	—	—	Normal men and women.

Abbreviations: CAD = coronary artery disease; P-E = prolate ellipsoid; SEE = standard error of the estimate.

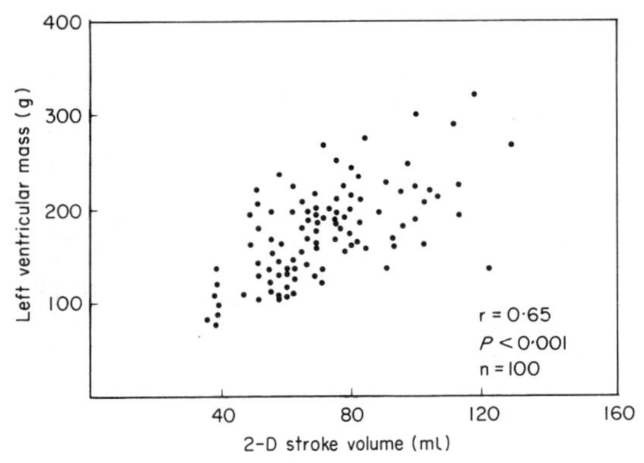

Figure 6 Relation between left ventricular mass measured from M-mode measurements by the Penn method (vertical axis) and stroke volume measured from two-dimensional echocardiographic left ventricular volumes calculated by the length-diameter prolate ellipsoid method (horizontal axis) in 100 normotensive or mildly hypertensive adults.

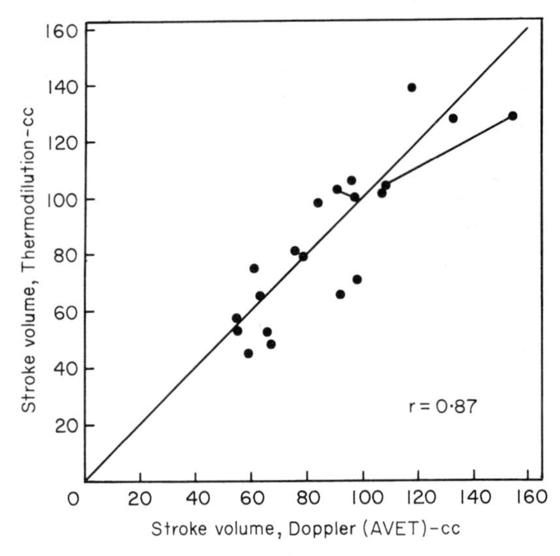

Figure 7 Stroke volume measurements by the aortic valve leading edge technique (AVET) and by thermodilution fall along the line of identity. The connected points represent two patients who had two separate measurements taken.

Doppler echocardiographic measurement of stroke volume and cardiac output

Doppler echocardiography has become an integral part of diagnostic echocardiography and, as is reviewed elsewhere in this issue, has been extensively applied to measure cardiac output. A discussion of echocardiographic evaluation of cardiac output would not be complete without at least brief mention of this modality, and thus a short comment on the strengths and shortcomings of this modality is included here.

Our group has performed a study comparing the relative accuracy of different combinations of methods of Doppler measurement and location where flow is measured[24]. Stroke volume measured using the aortic annular diameter (with the assumption that the aortic orifice is circular) and the spectral Doppler pattern of transaortic valve flow recorded in a 'five-chamber' view provided accurate results compared with invasive thermodilution (Fig. 7). The Doppler time-velocity integral was measured using the leading-edge (highest velocity) method. This result confirmed and extended a similar observation by Dittman *et al.*[53] Doppler stroke volume measured by the aortic annular leading-edge method correlated well with M-mode (Teichholz method) stroke volume in normotensive or hypertensive subjects studied in our laboratory (Fig. 1). Although Doppler stroke volume measurements have the advantage of being completely independent of assumptions about LV geometry, they have their own limitations: obtaining correct Doppler recordings is as at least as dependent on the skill of the operator as M-mode or 2-dimensional methods; and the problems of 'aliasing' or range ambiguity may, on occasion, cause problems in accurately measuring the velocity of rapid transaortic flow. However, adequate Doppler echocardiographic haemodynamic measurements can be obtained in a high percentage of subjects, making its application feasible in large-scale studies[54].

In summary, echocardiography provides three different modalities for the non-invasive measurement of stroke volume and cardiac output. M-mode methods have been most widely used and extensive data already exist relating cardiac output measurement by this method to cardiovascular effects of hypertension. This method is reasonably accurate in patients with symmetric ventricles, but cannot be used if LV shape or contraction are asymmetric. Fortunately, most hypertensive patients fall into the category of patients to which this method can be applied. Two-dimensional echocardiographic stroke volume and cardiac output measurement methods are theoretically attractive but suffer the limitation of having little data available validating their performance under clinical circumstances. However, it may be reasonable to extrapolate the numerous studies validating end-diastolic and end-systolic LV volume measurements by various 2-D echocardiographic methods to use of these methods for stroke volume measurements. However, 2-D methods systematically underestimate LV chamber volumes and stroke volume compared with invasively determined normal values. In contrast, Doppler echocardiographic measurements of blood flow across the aortic valve have been shown to correspond closely to invasively obtained stroke volumes. Results from our laboratory show virtually identical values for mean stroke volume by this Doppler method and the Teichholz M-mode method, with the former technique having the advantages of slightly lower standard errors of estimate compared with reference standards (about 10 ml vs 12–18 ml for stroke volume) and of applicability to patients with non-homogeneous ventricles.

We would like to thank Virginia Burns for her assistance in preparation of this manuscript.

References

[1] Savage DD, Garrison RJ, Kannel WB, Anderson SJ, Feinleib M, Castelli WP. Considerations in the use of echocardiography in epidemiology. The Framingham Study. Hypertension 1987; 9 (Suppl II): II40–4.
[2] Reichek N, Devereux RB. Left ventricular hypertrophy: Relationships of anatomic, echocardiographic and electrocardiographic findings. Circulation 1981; 63: 1391–8.
[3] Devereux RB, Casale PN, Wallerson DC *et al.* Cost-effectiveness of echocardiography and electrocardiography for detection of left ventricular hypertrophy in patients with systemic hypertension. Hypertension 1987; 9 (Suppl II): 69–76.
[4] Devereux RB. Evaluation of cardiac structure and function by echocardiography and other noninvasive techniques. In: Laragh JH, Brenner BM, eds. Hypertension: pathophysiology, diagnosis and treatment. New York: Raven Press 1990; 1479–92.
[5] Casale PN, Devereux RB, Milner M *et al.* Value of echocardiographic measurement of left ventricular mass in predicting cardiovascular morbid events in hypertensive men. Ann Intern Med 1986; 105: 173–8.
[6] Koren MJ, Casale PN, Savage DD, Laragh JH, Devereux RB. Relation of left ventricular mass and geometry to morbidity and mortality in men and women with essential hypertension. Ann Intern Med (in press).
[7] Levy D, Garrison RJ, Savage DD, Kannel WB, Castelli WP. Left ventricular mass and incidence of coronary heart disease in an elderly cohort. Ann Intern Med 1989; 110: 101–7.
[8] Levy D, Garrison RJ, Savage DD, Kannel WB, Castelli WP. Prognostic implications of echocardiographically determined left ventricular mass in the Framingham Heart Study. N Engl J Med 1990; 322: 1561–6.
[9] Koren MJ, Savage DD, Laragh JH, Devereux RB. Relation of hemodynamic parameters to prognosis in essential hypertension. Am J Hypertension 1990; 3: 13A (Abstr).
[10] Ganau A, Devereux RB, Pickering TG *et al.* Relation of left ventricular hemodynamic load and contractile performance to left ventricular mass in hypertension. Circulation 1990; 81: 25–36.
[11] de Simone G, Volpe M, Wallerson DC, Devereux RB, Camargo MJF, Laragh JH. Relation of ventricular geometry to hemodynamic load in Goldblatt rats. Am J Hypertension 1989; 2: 64A.
[12] Sahn D, DeMaria A, Kisslo J, Weyman A. The Committee on M-mode standardization of the American Society of Echocardiography: Recommendations regarding quantitation in M-mode echocardiography: Results of a survey of echocardiographic measurements. Circulation 1978; 58: 1072–83.
[13] Popp RL, Harrison DC. Ultrasonic cardiac echography for determining stroke volume and valvular regurgitation. Circulation 1970; 41: 493–502.
[14] Pombo JF, Troy BL, Russel RO. Left ventricular volumes and ejection fraction by echocardiography. Circulation 1971; 43: 480–90.
[15] ten Cate FJ, Kloster FE, van Dorp WB, Meester GI, Roelandt J. Dimensions and volumes of left atrium and ventricle determined by single beam echocardiography. Br Heart J 1974; 36: 737–46.
[16] Kronik G, Slany J, Mosslacher H. Comparative value of eight M-mode echocardiographic formulas for determining left ventricular stroke volume. A correlative study with thermodilution and left ventricular single-plane cineangiography. Circulation 1979; 60: 1308–16.
[17] Feigenbaum H. New aspects of echocardiography. Circulation 1973; 47: 833–42.
[18] Gibson DG. Estimation of left ventricular size by echocardiography. Br Heart J 1973; 35: 128–34.
[19] Belenkie I, Nutter DO, Clark DW, McCraw DB, Raizner AE. Assessment of left ventricular dimensions and function by echocardiography. Am J Cardiol 1973; 31: 755–62.
[20] Ludbrook P, Karlinger JS, Peterson K, Leopold G, O'Rourke RA. Comparison of ultrasound and cineangiographic measurements of left ventricular performance in patients with and without wall motion abnormalities. Br Heart J 1973; 35: 1026–32.
[21] Teichholz LE, Kreulen T, Herman MV, Gorlin R. Problems in echocardiographic determinations: Echocardiographic-angiographic correlations in the presence or absence of asynergy. Am J Cardiol 1976; 37: 7–11.
[22] Kiowski W, Randall OS, Steffens TG, Julius S. Reliability of echocardiography in assessing cardiac output. A comparison study with a dye dilution technique. Klin Wochenschr 1981; 59: 1115–20.
[23] Hutchins GM, Bulkley BH, Moore GW, Piasio MA, Lohr FT. Shape of human cardiac ventricles. Am J Cardiol 1978; 41: 646–54.
[24] Dubin J, Wallerson DC, Cody RJ, Devereux RB. Comparative accuracy of Doppler echocardiographic methods for clinical stroke volume determination. Am Heart J 1990; 120: 116–23.
[25] Popp RL, Alderman EL, Brown OR, Harrison DC. Sources of error in calculation of left ventricular volume by echocardiography. Am J Cardiol 1973; 31: 152A.
[26] Gehrke J, Leeman S, Raphael M, Pridie RB. Non-invasive left ventricular volume determination by two-dimensional echocardiography. Br Heart J 1975; 37: 911–6.
[27] Eaton LW, Maughan WL, Shoukas AA, Weiss JL. Accurate volume determination in the isolated ejecting

canine left ventricle by two-dimensional echocardiography. Circulation 1979; 60: 320–6.

[28] Schiller NB, Acquatella H, Ports TA *et al.* Left ventricular volume from paired biplane two-dimensional echocardiography. Circulation 1979; 60: 547–55.

[29] Folland ED, Parisi AF, Moynihan PF, Jones DR, Feldman CL, Tow DE. Assessment of left ventricular ejection fraction and volumes by real time, two-dimensional echocardiography. Circulation 1979; 60: 760–6.

[30] Wyatt HL, Heng MK, Meerbaum S *et al.* Analysis of mathematic models for quantifying volume of the formalin-fixed left ventricle. Circulation 1980; 61: 1119–25.

[31] Silverman NH, Ports TA, Snider AR, Schiller NB, Garrison E, Heilbron DC. Determination of left ventricular volume in children: Echocardiographic and angiographic comparisons. Circulation 1980; 62: 548–57.

[32] Erbel R, Schweizer P, Lambertz H *et al.* Echoventriculography. A simultaneous analysis of two-dimensional echocardiography and cineventriculography. Circulation 1983; 67: 205–15.

[33] Wyatt HL, Heng MK, Meerbaum S *et al.* Cross sectional echocardiography. 1. Analysis of mathematic models for quantifying mass of the left ventricle in dogs. Circulation 1979; 60: 1105–13.

[34] Reichek N, Helak J , Plappert T, St John Sutton M, Webber KT. Anatomic validation of left ventricular mass estimates from clinical two-dimensional echocardiography: initial results. Circulation 1983; 67: 348–52.

[35] Stack RS, Ramage JE, Bayman RP, Rembert JC, Phillips MD, Kisslo JA. Validation of in vivo two-dimensional echocardiographic dimension measurements using myocardial mass estimates in dogs. Am Heart J 1987; 113: 725–31.

[36] Erbel R, Schweizer P, Hevin G, Mayer J, Effert S. Apical two-dimensional echocardiography: Normal values for single and bi-plane determination of left ventricular volume and ejection fraction. Dtsch Med Wochenschr 1982; 107: 1872–7.

[37] Gordon EP, Schnittger I, Fitzgerald PJ, Williams P, Popp RL. Reproducibility of left ventricular volumes by two-dimensional echocardiography. J Am Coll Cardiol 1983; 2: 506–13.

[38] Byrd BP III, Wahr D, Wang YS, Bouchard A, Schiller NB. Left ventricular mass and volume/mass ratio determined by two-dimensional echocardiography in normal adults. J Am Coll Cardiol 1985; 6: 1021–5.

[39] Wahr DW, Wang YS, Schiller NB. Left ventricular volumes determined by two-dimensional echocardiography in a normal adult population. J Am Coll Cardiol 1983; 3: 863–8.

[40] Helak JW, Reichek N. Quantitation of human left ventricular mass and volume by two-dimensional echocardiography: In vitro anatomic validation. Circulation 1981; 63: 1398–407.

[41] Pini R, Ferrucci L, DiBari M *et al.* Two-dimensional echocardiographic imaging: In vitro comparisons of conventional and dynamically focused anular array transducers. Ultrasound Med Biol 1987; 13: 643–50.

[42] Pini R, Monnini E, Masotti L *et al.* 3-D Imaging in Medicine: Algorithms, systems, applications. In: Fuchs H, Hoehne KH, Pizer SM, eds. Berlin: Springer-Verlag 1990; 262–74.

[43] Harrison MR, King DL, King DL Jr, Smith MB, Kwan OL, DeMaria AN. Ultrasound beam orientation during standardized imaging: Assessment by 3-dimensional echocardiography. J Am Soc Echocardiography 1990; 3: 228A.

[44] Katz AS, Wallerson DC, Pini R, Devereux RB. Visually determined long-axis and short-axis parasternal views and four chamber apical views do not represent paired orthogonal projections. J Am Coll Cardiol 1990; 15: 94A.

[45] Himelman RB, Cassidy MM, Landzberg JS, Schiller NB. Reproducibility of quantitative two-dimensional echocardiography. Am Heart J 1988; 115: 425–31.

[46] Devereux RB, Reichek N. Echocardiographic determination of left ventricular mass in man: anatomic validation of the method. Circulation 1977; 55: 613–8.

[47] Devereux RB, Alonzo DR, Wtas EM *et al.* Echocardiographic assessment of left ventricular hypertrophy: Comparison to necropsy findings. Am J Cardiol 1986; 57: 450–8.

[48] Julius S, Conway J. Hemodynamic studies in patients with borderline blood pressure elevation. Circulation 1968; 38: 282–8.

[49] Messerli FH, Ventura HO, Reisen E *et al.* Borderline hypertension and obesity: Two prehypertensive states with elevated cardiac output. Circulation 1982; 66: 55–60.

[50] Hammond IW, Devereux RB, Alderman MH *et al.* The prevalence and correlates of echocardiographic left ventricular hypertrophy among employed patients with uncomplicated hypertension. J Am Coll Cardiol 1986; 7: 639–650.

[51] Ibrahim MM, Tarazi RC, Dustan HI, Bravo EL, Gifford RW. Hyperkinetic heart in severe hypertension: A separate clinical hemodynamic entity. Am J Cardiol, 1975; 35: 667–74.

[52] Finkelman S, Worcel M, Agnest A. Hemodynamic patterns in essential hypertension. Circulation 1965; 31: 351–68.

[53] Dittman H, Voelker W, Karsch KR, Seipel L. Influence of sampling site and flow area on cardiac output measurements by Doppler echocardiography. J Am Coll Cardiol 1987; 10: 818–23.

[54] Julius S, Jamerson K, Mejia A, Krause L, Schork N, Jones K. The association of borderline hypertension with target organ changes and higher coronary risk. Tecumseh Blood Pressure Study. J Am Med Assoc 1990; 264: 354–8.

Impedance cardiography for cardiac output measurement: An evaluation of accuracy and limitations

S. W. WHITE*†‡, A. W. QUAIL*‡, P. W. DE LEEUW§, F. M. TRAUGOTT*, W. J. BROWN*, W. L. PORGES¶ and D. B. COTTEE*‡

*Discipline of Human Physiology and Neuroscience Group, Faculty of Medicine, University of Newcastle, †Departments of Cardiology, and ‡Anaesthesia and Intensive Care, Royal Newcastle Hospital, and ¶Department of Veterinary Clinical Studies, University of Sydney, Australia, and §Department of Medicine, Zuiderziekenhuis, Rotterdam, The Netherlands

KEY WORDS: Transthoracic impedance cardiography, stroke volume, cardiac output, hypertension, exercise, blood resistivity.

The Kubicek thoracic cylinder model of impedance cardiography (IC) for measuring beat-by-beat stroke volume (SV) was evaluated in controlled studies using the electromagnetic flowmeter (FM) as the reference technique. Assuming the validity of the Kubicek equation for stroke volume calculation, IC stroke volume was found to be a linear function of EM values at any one haematocrit over a wide range of SV, but the slope of the relationship fell as haematocrit fell. Experiments using the same equation in dogs, in which blood resistivity in vivo (ρ_τ) was made the dependent variable, and the EM-derived value was used for stroke volume, showed that ρ_τ was almost constant over a wide range of haematocrits. These findings were supported by studies in man and rabbit where Fick and thermodilution-derived values were used for stroke volume. When these data were applied to normotensive and hypertensive human subjects with normal hearts and lungs in controlled studies at rest, during tilting, with drug therapy and on exercise, IC measured stroke volume and cardiac output with a variability at least as good as the 9–11% acceptable for clinical use. This conclusion applied to thoracic configurations of different sizes and shapes from adult man to the neonate. In chronic disease states, while assessments of relative changes are valuable, absolute data are questionable. Further research is required under these conditions, as it is also for other models of IC, which are based on different assumptions.

Introduction

The early impedance work directed to non-invasive measurement of human function[1–4] received a new stimulus when the call came for measuring cardiac function in space[5,6]. However, Kubicek and colleagues[6,7] found that the theoretical basis for using the thorax as an analogue cylinder from which a stroke volume was ejected was unworkable. Nevertheless, using an empirical correction to the formula, based on an estimate of the impedance change (ΔZ) which would have obtained if other movements of blood in and out of the thorax had not occurred, Patterson in his thesis[7] (Fig. 1b) found a good agreement in man between his empirically derived measurements of cardiac output and those calculated using the Fick approach (Fig. 2) in patients without vascular shunts.

The original formula based on theoretical considerations was modified by substituting $T(dZ/dt_{max})$ for ΔZ i.e.

$$SV = \frac{\rho . L^2}{Z_0^2} . \Delta Z \quad \text{became} \quad SV = \frac{\rho . L^2}{Z_0^2} . T(dZ/dt_{max})$$

where SV = stroke volume, ml; ρ = blood resistivity (Ωcm); L = mean midline distance between ventral and dorsal recording electrodes (cm), Z_0 = mean thoracic impedance (Ω), T = ejection time as determined by the time between intercept of the baseline with the onset of a decrease of impedance preceding its maximal decrease during systole, and the commencement of the 2nd heart sound (s), dZ/dt_{max} = the maximum rate of change of impedance during ventricular ejection ($\Omega\, s^{-1}$).

Since that time, there has been an increasing number of papers published concerning the variable accuracy of the method under a variety of

Supported by the National Health and Medical Research Council of Australia, and ICI Pharmaceuticals, ICI Australia Operations Pty Ltd, ICI Australia.

Address for correspondence: Professor Saxon White, Discipline of Human Physiology, University of Newcastle, New South Wales 2308, Australia.

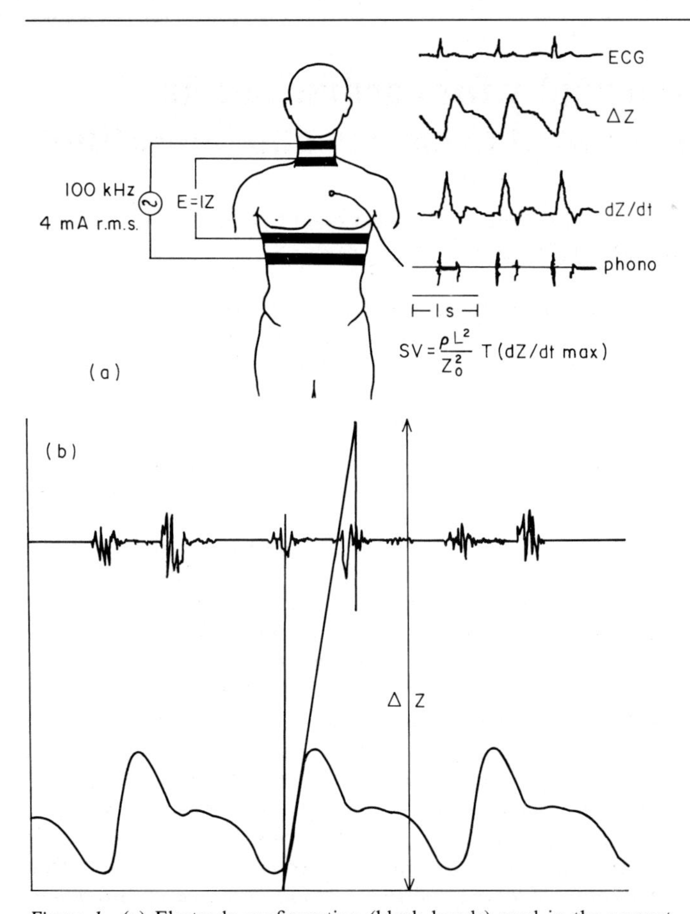

Figure 1 (a) Electrode configuration (black bands) used in the present study. ECG = electrocardiogram; ΔZ = change in impedance recorded from inner electrodes with each cardiac cycle; dZ/dt_{max} = differentiated ΔZ signal; phono = phonocardiograph. (b) Enhanced trace redrawn from Patterson's thesis[7] with permission. Predicted ΔZ is graphically determined by extrapolating the maximum decreasing slope of the impedance change waveform from the base of the curve at the start of systole to the end of ventricular systole as indicated by the second heart sound. This principle has been followed in the present evaluation studies, but the absolute criteria for the calculation of stroke volume have been changed in minor ways, and are described in the text. Following convention (see also (a) the ΔZ record as shown is inverted. Upper tracing = heart sounds; lower tracing = ΔZ record with each cardiac contraction.

conditions in health and disease. Analysis of papers up to 1980[8] suggested that a major reason for this state of affairs was the lack of accurate independent methods in the clinical departments where comparisons were made. Another reason for concern was the failure to test rigorously the empirical method for measuring cardiac function over the full range of physiological performance, e.g. over a wide range of heart rates, stroke volumes, inotropic states, haematocrits, thoracic sizes and so on. A further point was the failure to test the instrument using normal hearts and lungs before going on to study pathophysiological states.

Methodological considerations

Paired impedance electrodes (aluminium mylar, adhesive strip electrodes; Instrumentation for Medicine Inc., Conn. 06832) were positioned circumferentially 2–3 cm apart on the shaved skin of the neck and lower thorax (Fig. 1a, Fig. 3). The outer two electrodes were connected to a constant

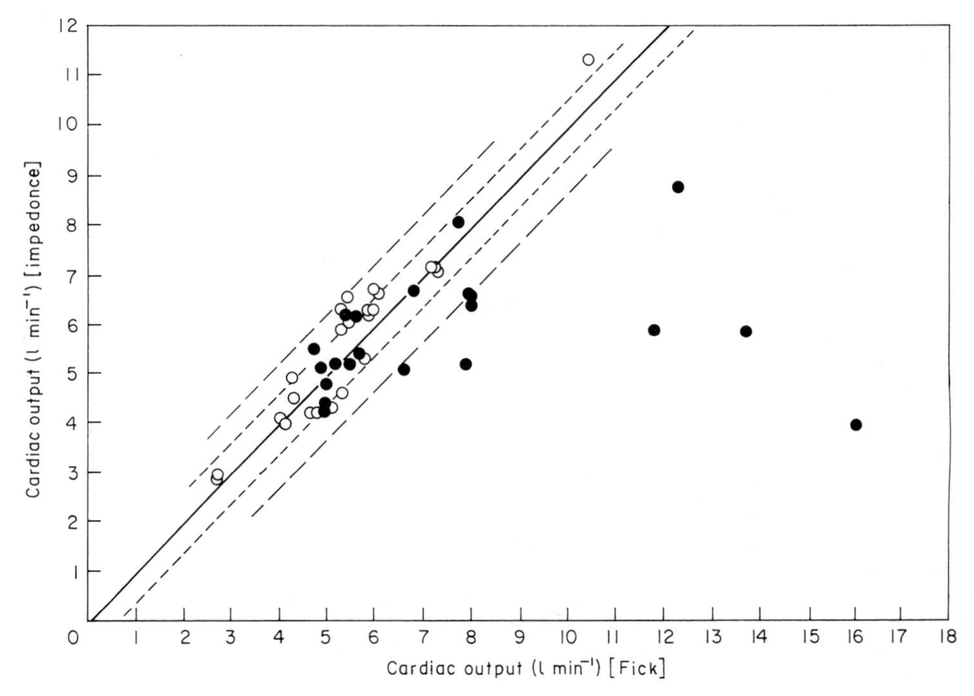

Figure 2 Comparison of two cardiac output measurement techniques, direct Fick (CO_{Fick}), and impedance (CO_Z); data from Patterson's thesis from [7], with permission. ● = intracardiac shunts (atrial and septal defects). ○ = data from normal hearts.

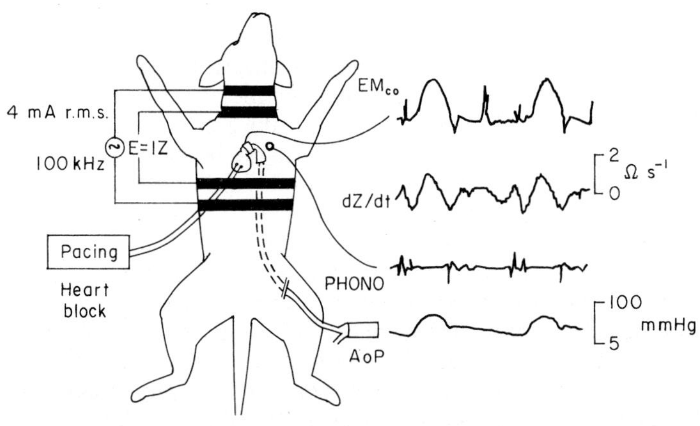

Figure 3 The impedance cardiograph was evaluated in conscious dogs in complete experimental heart block EM_{CO} = electromagnetic flowmeter signal, dZ/dt = impedance cardiograph signal; PHONO = phonocardiograph; AoP = aortic pressure. From [17] with permission.

current source (4 mA, 100 kHz) and the inner two (positioned routinely at the base of the neck, and at the xiphisternal joint) were used to record the ECG, basic impedance of the thorax (Z_0), and also the small impedance change that occurs during the cardiac cycle (ΔZ) plus its first derivative (dZ/dt_{max}). Heart sounds were monitored using a contact sensor (Hewlett-Packard 21050A). All signals were monitored on a chart recorder with a high frequency response to ensure, in particular, a faithful recording of the peak dZ/dt_{max} signal. Measurements at rest were made in functional residual capacity, to

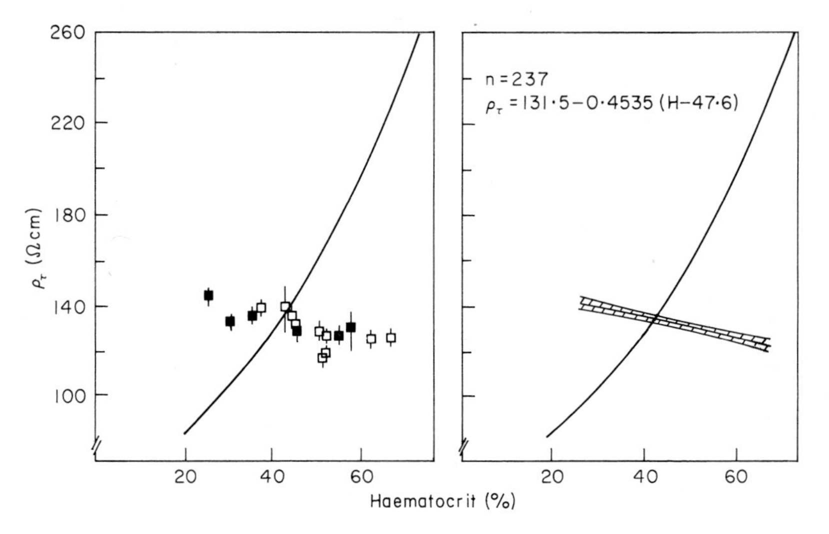

Figure 4 Relationship between thoracic resistivity (ρ_τ) and haematocrit in the dog. Left: closed squares represent mean ρ_τ in four dogs in which haematocrit was altered by haemorrhage and infusion; open squares represent mean ρ_τ in remaining three dogs in which haematocrit was unaltered; bars represent ±SEM. Right: pooled linear regression line for 237 paired observations in all seven dogs; hatched areas represent 1 SD above and below mean. In vivo findings of present study are contrasted with those bench-derived values of Geddes and Sadler[9], who found an exponential relationship of increasing blood resistivity as haematocrit increased (see exponential curve each panel above). From [14] with permission.

eliminate transient effects of respiration on ΔZ, and thus dZ/dt_{max} (see below).

Analysis of the limitations of impedance cardiography

Our own evaluative experience with impedance cardiography (IC) has been entirely with the Instrumentation for Medicine (IFM), Minnesota, Model 400. In our studies[8, 14, 15, 17], using the criteria for the measurement and underlying assumptions suggested by Kubicek and colleagues[6], we decided in the first instance to examine the normal circulation, and thereby define the relationship, be it linear or curvilinear, between impedance-derived stroke volume measurement and the most carefully controlled independent method available. We chose the electromagnetic method of stroke volume measurement as a reference, and used large, fit dogs with probes mounted at prior operations on the aortic root some 10 days before experiments (Fig. 3). At operation, we also put the animals into complete heart block and controlled their ventricular rates with a pacemaker unit; therefore , by variable pacing at rest we could alter stroke volume over the physiological range and beyond.

But there was a further problem, although we were encouraged by the success of the empirical adjustments made by Kubicek, Patterson *et al.*, we were less confident about the suggested blood resistivity (ρ) correction for the prevailing haematocrit (Hct). This was because the bench-derived curves for ρ-Hct relationships[9–11] may not hold in vivo. Blood resistivity is known to be a function of, for example, red cell velocity, as well as Hct. These factors might well interact, for example, in the case of patients under intensive care receiving infusions, or in the case of exercising (anaemic) people, when the ejection velocity of red cells during the high inotropic state of systole may modify in unknown ways the ρ-Hct bench-derived curve. We also noted from the literature that when IC was compared with a standard technique during exercise[12], for example, or when infusions were given[13], that the errors were systematically high, or low, when the ρ-Hct bench-derived curves were used in the presence of high or low Hct. Moreover, the data overall implied IC was linear at any one ρ value.

For the above reasons, we decided to investigate the in vivo blood resistivity/haematocrit relationship first[8,14], and to do this we rearranged the

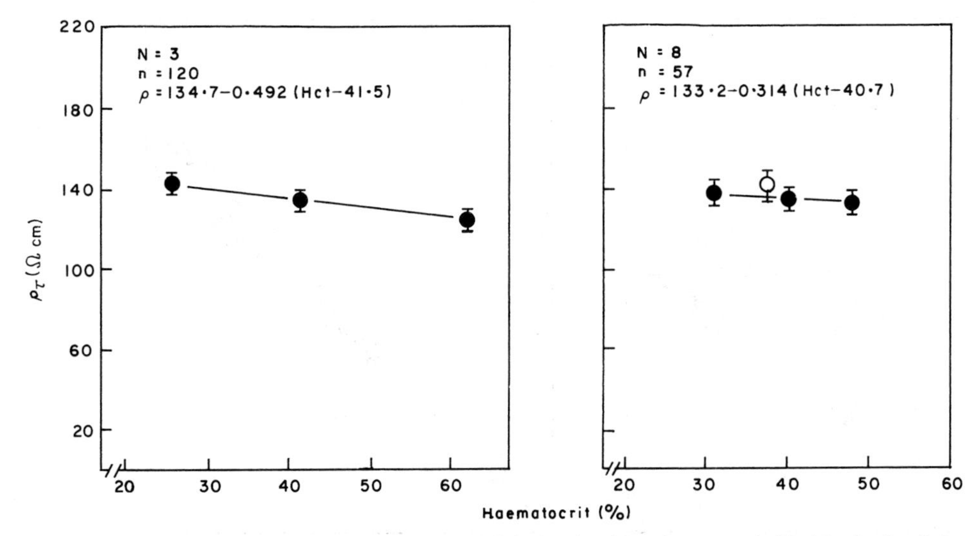

Figure 5 Linear regression lines relating blood resistivity in vivo (ρ) to haematocrit (Hct) in the dog (left panel) and in man (right panel). Open circle (right panel) represents mean ρ value at a mean haematocrit of 37·4% in five rabbits. Symbol represents one standard error above and below the mean. From [15] with permission.

Kubicek formula and made blood resistivity the dependent variable:

$$\rho = (SV.Z_0^2)/(L^2 dZ/dt_{max}.T)$$

Stroke volume was measured by carefully calibrated electromagnetic flowmeters, and haematocrit was varied by haemorrhage and simultaneous replacement infusion of Dextran 70 in normal saline. In contrast to the direct and exponential bench ρ-Hct relationship, thoracic resistivity (ρ_τ, as we termed it) was almost constant at 135 Ω cm over a range of haematocrit of half-normal to 150% of normal. In fact, the in vivo ρ_τ-Hct relationship (Fig. 4) was linear and slightly inverse, but by no more than +6·3 Ωcm at Hct 26%, and −11·8 Ωcm at Hct 66% about a mean ρ_τ of 135±1·0 Ωcm at Hct 40%. Use of a constant value of 135 Ωcm for ρ_τ in the Kubicek formula would therefore result in a maximum error of less than 10% in stroke volume calculation over a wide range of haematocrits, and the size of this error was not raised by physiological extremes of heart rate, or of myocardial inotropic state created by infused isoprenaline[14]. We did detect a tendency for non-linearity at heart rates below 60 beats min^{-1} [8], but it would be rare to encounter such low heart rates under natural circumstances. Other less controlled studies in man and rabbit showed a similar ρ_τ-Hct relationship (Fig. 5), which led us to postulate[15] that the use of a constant ρ_τ would hold despite the varying thoracic size and shape in many mammalian species, probably because of the common functional (vascular) architecture within. This appears to be confirmed by Gotshall's[16] studies e.g. in the 1-day-old rat, in which stroke volume is as small as 0·03 ml.

When ρ_τ was used in subsequent studies in man, dog and rabbit[15,17], the relationship between IC and independent measures was vastly improved. This is seen at its best in the retention of a good agreement *despite* variations in haematocrit when simultaneous beat-by-beat comparisons are made (Fig. 6) in the dog when the standard reference was the electromagnetic flowmeter. The greater scatter (Fig. 7) seen when less control can be exerted over the independent method (e.g. thermodilution in man) can be attributed in part to the variation introduced, for example, by non-instantaneous mixing of indicator (negative heat) due to the prolonged injection time, untoward heat losses, poor thermistor response times and the interactive effects of injection time and the respiratory cycle. Fick and indicator dilution methods are 'averaging' techniques, and comparisons with beat-by-beat methods are at risk unless extraordinary experimental control is used.

A different limitation on cardiac output measurement is unique to IC. All our evaluative work was carried out at rest, using samples of five consistent dZ/dt wave-form heights at *functional residual capacity (FRC)*. During exercise in man, however,

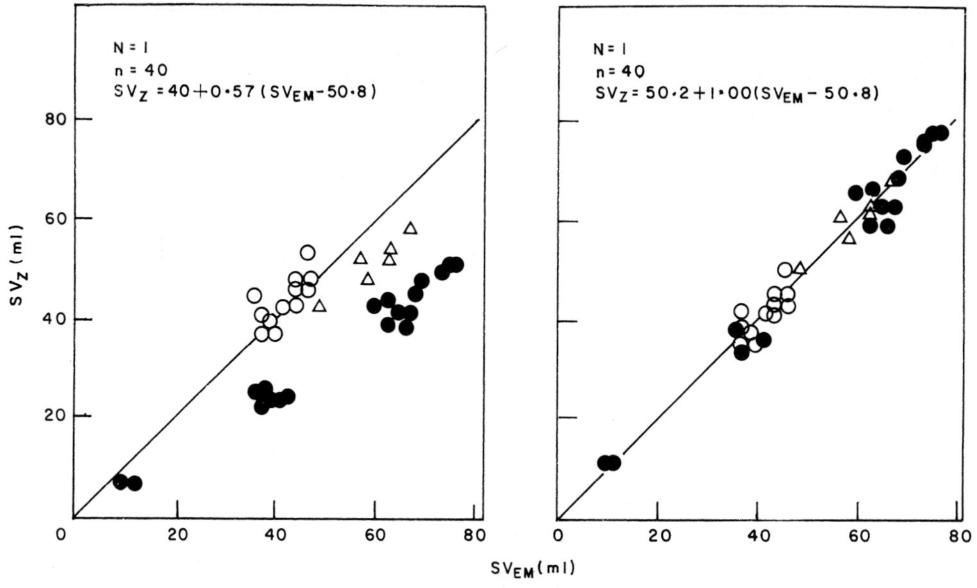

Figure 6 Relationship between simultaneous measurements of impedance (SV_Z) and electromagnetic flow stroke volumes (SV_{EM}) in one dog at three different haematocrits; open circles 44·5%, open triangles 36·0%, closed circles 26·2%. Left panel: data calculated using ρ from Geddes and Sadler (1973) relationship (r = 0·758). Right panel: the same data calculated using in vivo ρ_τ experimentally derived from present study (r = 0·983). Straight line represents line of equal value. From [15] with permission.

we noted that if SV measurements were made at FRC (particularly when subjects 'helpfully' forcibly exhale), we may bias downwards the average SV due to suppression of venous return. This effect may be greater during exercise than at rest and lead to cardiac output values less than those derived from techniques such as Fick and indicator dilution, which measure an average cardiac output without patients' suppressing respiration in expiration. A typical analysis of SV measurement during the extremes of the respiratory cycle is shown in Table 1. At rest, the SV measured varies by some 25% between FRC and near inspiratory capacity, and by 50% during moderate exercise. Since the *average* SV measured by either the indicator or Fick method lies about halfway between, the IC value at rest is likely to be lower by 10–12% than the equivalent measurement made by an averaging technique; at moderate exercise the IC value would be some 25% lower. These conclusions may explain underreading discrepancies for IC during exercise, when measurements are made at FRC. For this reason, to ensure measurements are free of any concomitant *transient* respiratory effects on venous return, ΔZ and hence dZ/dt_{max}, we now train our subjects during exercise to cease breathing in mid-inspiratory cycle, rather that at FRC.

Aspects of accuracy

The results pertaining to accuracy from our own laboratories are shown in Fig. 7[17]. In the dog (top left, Fig. 7) the stroke volume varied from 9 to 76 ml about a mean SV_Z value of 50·8 ± 2·58 (SEM) ml, i.e. ±5% of the mean value. In a separately reported study in three other dogs, similar comparisons[8] in each dog did not differ significantly from the line of identity, and the pooled data for 105 comparisons showed a linear relationship over a range of 8–46 ml. Eighty-two per cent of the points fell within ±20% of the line of equal value, and the standard error of the estimate was ±0·36 ml about a mean SV_Z of 22·1 ml, i.e. ±2% of the mean value. These excellent figures for accuracy must be compared with the less impressive results of studies in man (Fig. 7, left lower panel) and rabbit (Fig. 7, right upper panel) in our hands where the accuracy of beat-by-beat SV_Z measurements were compared simultaneously with different independent 'averaging' techniques; direct Fick and thermodilution, and Fick techniques, respectively. In all cases SV_Z was calculated using the Kubicek formula and ρ_τ. In neither case (because of clinical circumstances in man, and other circumstances in the anaesthetized rabbit) was ventilation controlled or stroke volume

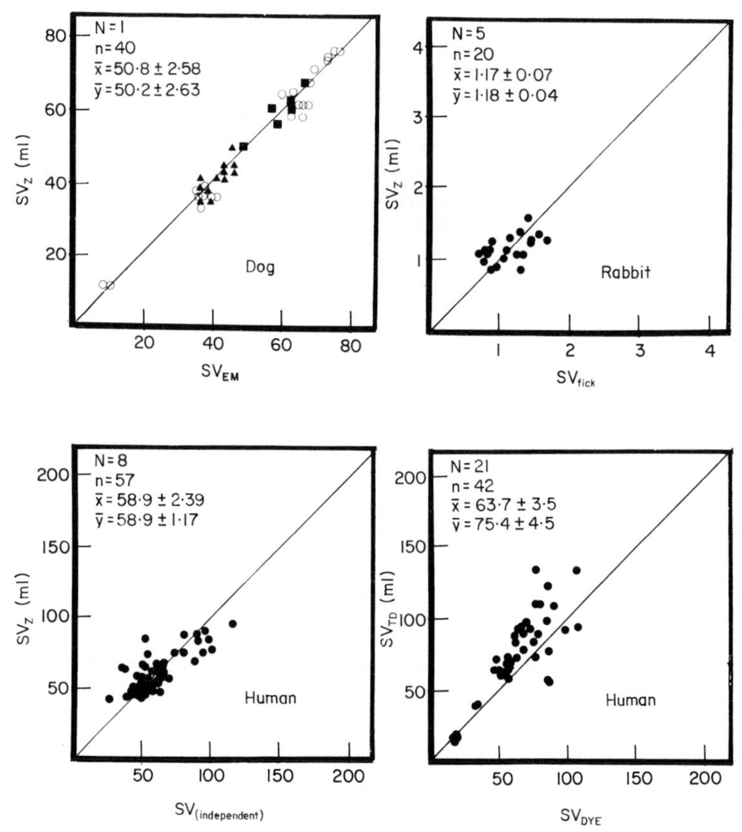

Figure 7 Comparison between different methods of measuring stroke volume in different species, i.e. dog (upper left panel), rabbit (upper right panel) and in humans (lower panels). SV_Z = impedance stroke volume, SV_{TD} = thermal dilution stroke volume, SV_{EM} = electromagnetic flowmeter stroke volume, SV_{fick} = Fick principle stroke volume, SV independent = stroke volume measured using either Fick or thermodilution methods, SV_{dye} = dye dilution stroke volume. Symbols in upper left panel represent stroke volumes at different haematocrits. ○ = 26%, ■ = 36%, △ = 45%. From [17] with permission.

experimentally varied. There was nonetheless a good agreement between means but, due to the data scatter, correlation coefficients (0·82 for man, 0·47 for rabbit) were much less than for the carefully controlled, beat-by-beat comparison in the dog. Finally, in 42 paired comparisons in 21 human patients at rest (Fig. 7, lower right panel), simultaneous measurements were made of SV calculated from thermodilution and dye dilution curves. A better spread of spontaneous stroke volumes was obtained, there was a significant regression coefficient (0·823), the calculated regression line was not shifted from the line of identity and the standard error of the estimate of the pooled data was 4·3% (±3·3 ml) of the mean thermodilution stroke volume value of 74·5 ml. These data show similar scatter to that with many other IC comparisons with standards in the literature. They suggest to us that, in our hands, using a multitude of cardiac output measuring methods the accuracy of SV measured by IC in man and smaller animals compares favourably with the clinical usage of thermodilution, and that the reproducibility of a single IC SV measurement is better than that obtained using thermodilution.

Precision and reproducibility

In acquiring insight into the use, limitations and accuracy of IC, we measured intra- and inter-observer precision in assessing the data wave forms. Table 2 shows how well a single *trained* observer can read six different stroke volume complexes when

Table 1 Effect of measuring stroke volume (SV) and heart rate (HR) at different phases of the respiratory cycle, on the determination of cardiac output (CO) at rest, and during moderate exercise. For resting values, N = 12 separate determinations during each phase of the respiratory cycle, in one subject. For exercise values, N = 10 separate determinations during each phase in the same subject

	Expiration (Functional residual capacity)			Inspiration (Near inspiratory capacity)		
	SV (ml)	HR (min^{-1})	CO ($l\ min^{-1}$)	SV (ml)	HR (min^{-1})	CO ($l\ min^{-1}$)
Rest						
Mean	46	74	3·4	59	77	4·5
SEM	1·3	2·3	0·08	3·3	5·1	0·13
% expiration measurements	100	100	100	128	104	132
Exercise						
Mean	42	134	5·6	63	134	8·4
SEM	3·4	3·6	0·55	2·2	4·4	0·26
% expiration measurements	100	100	100	151	100	150

Table 2 Observer precision

Observed sequence	1	2	3	4	5
Mean SV (ml) N=6	68·4	67·5	69·2	68·3	67·8
Grand Mean N=30			68·2		
% Grand Mean	−0·3	−1·1	+1·4	+0·1	−0·7
Max deviation			<2·5%		

Table 3 Interobserver precision

Observer	WB experienced	AB learning	CC new
Mean SV (ml) N=6	68·3	74·3	67·1
% experienced observer	100	109	98
Probability	—	<0·01	>0·05

complexes are presented in random order so that each complex is read five times. After breaking the code and retabulating the results, analysis of variance indicates no difference between observation sequence, and shows that intraobserver precision is sustained within 2·5% of the mean value of 68·2 ml.

In a second study (Table 3), we examined interobserver precision using an experienced observer, a learning observer, and a new observer shown once by the experienced observer how the records should be read. Each observer read the same six complexes once. Analysis of variance of the data showed that the learning observer overread the complexes significantly by 9%, and that this was entirely due to errors in the estimate of T (ejection time). Small differences in reading T have been consistently observed in our laboratory to cause the greatest errors in stroke volume estimates. The new observer had been trained well, if briefly. We did not measure the learning curve for reading as did Wong *et al.*[18] when they showed some nine trials were necessary to train observers to the precision of an experienced observer. Nevertheless, all the data show that the records are easy to assess, require relatively simple training, and can be read with reproducible precision to less than 3% of the true value.

A further study was devised to evaluate the reproducibility of measuring cardiac output in man when a series of measurements were made several weeks apart (Table 4). Six normotensive non-athletic males (matched for age, weight and height) were examined on four separate occasions at 1-week intervals. The study consisted of an exercise schedule using an Elema 380B cycle ergometer, with 3-min incremental steps in workload up to moderate exercise levels (900 kpm). Cardiac output was determined using the impedance cardiograph (IFM 400), with a single observer making all measurements.

Table 4 Mean cardiac outputs ($l\,min^{-1}$) ($\pm$SEM) for six normotensive subjects at rest and during graded exercise

Time	Control 1	Control 2	300 kpm	450 kpm	600 kpm	750 kpm	900 kpm
Week 1	4·7 ±0·48	4·4 ±0·30	7·1 ±0·75	8·9 ±0·69	10·2 ±0·89	11·1 ±0·77	13·5 ±1·03
Week 2	4·4 ±0·44	4·2 ±0·17	7·8 ±0·61	9·7 ±0·92	11·7 ±1·16	11·6 ±0·75	13·1 ±1·29
Week 3	5·1 ±0·44	4·5 ±0·25	8·2 ±0·59	10·2 ±0·58	11·2 ±0·70	12·4 ±0·67	13·6 ±1·34
Week 4	4·2 ±0·29	4·1 ±0·18	7·8 ±0·57	9·8 ±0·48	10·8 ±0·53	12·4 ±0·66	14·5 ±1·25

Age = 37 ± 2·3 years; weight = 74 ± 2·2 kg.

There was no significant difference in measurements across weeks using two-way ANOVA within subjects, either at rest, or during exercise, but there was a highly significant rise in cardiac output as the subjects exercised, as would be expected. There was no significant interaction effect of time and the exercise cardiac output response as might occur if there was an effect of training on the exercise capability of the subject.

Clinical applications — studies in hypertension

Following this initial study of reproducibility in upright exercise, experiments using normotensive and essential hypertensive male subjects were carried out (Table 5), this time using two preparations of β-blocker (atenolol) and two placebos, with the four treatments given in random order to the same male subjects. The time course and upright exercise schedule was otherwise the same as in the initial study, and was undertaken in a normotensive group (N = 6), and in a newly diagnosed, untreated essential moderately hypertensive (N = 5) group, matched with the initial study-group for age, weight and height. There were no significant differences in cardiac outputs during exercise between the placebo responses either at rest, or during exercise, in either the normotensive or the hypertensive group. These data confirm the interpretation of good reproducibility of cardiac output measurement, but in the hypertensive cardiac state as well as in the normotensive situation. However, the data concerning the normotensive group show a clear-cut and similar cardiac output reduction due to atenolol treatment at rest and during exercise. By contrast, this atenolol effect was not present in the hypertensive group at rest or early during exercise, but a significant interaction of treatment and exercise indicated a progressively reduced exercise response as external work output increases in the group of hypertensives. Other impedance cardiography studies were carried out to evaluate haemodynamic abnormalities in a larger group of treated essential hypertensive patients with high blood pressure. In 18 subjects, in whom treatment had been withdrawn at least 3 weeks before study, simultaneous measurements of cardiac output by the impedance and by the dye dilution technique were carried out while these individuals were supine after a night's rest. For calculation of stroke volume by IC, ρ was again taken as 135 Ωcm. Cardiac output as determined by dye dilution averaged $4{\cdot}7 \pm 0{\cdot}43$ (SEM) $l\,min^{-1}$, with a corresponding figure of $4{\cdot}8 \pm 0{\cdot}36\,l\,min^{-1}$ for the impedance data. Even though there was close agreement between the means, and the correlation coefficient exceeded 0·80, the data analysed in this way hid some important information. When we plotted our results more appropriately, as suggested by Bland and Altman[19], the picture shown in Fig. 8 emerged. On average, there was, indeed, only a very small difference between the results obtained by both methods. Nevertheless, there was considerable scatter in the differences of individual measurements. In fact these differences varied from $-1{\cdot}96$ to $+3{\cdot}10\,l\,min^{-1}$. Interestingly, the agreement between the two methods was closest at the extremes of cardiac output. However, we note (as mentioned earlier) in interpreting these data that it is difficult to justify a comparison between an averaging technique and one that measures stroke volume on a beat-by-beat basis.

In another study, we assessed the reproducibility of impedance cardiac output measurements in 21 hypertensive patients in hospital on 2 different days,

Table 5 Mean cardiac outputs ($l\,min^{-1}$) ($\pm$SEM) for six normotensive and five hypertensive subjects at rest and during graded exercise with placebo or β-blocker treatment

Treatment	Control 1	Control 2	300 kpm	450 kpm	600 kpm	750 kpm	900 kpm
			*Normotensives**				
atenolol	4·0 ±0·52	3·8 ±0·52	5·9 ±0·55	7·4 ±1·09	8·8 ±1·11	10·2 ±1·24	11·1 ±1·39
atenolol+ chlorthal.	4·2 ±0·41	3·9 ±0·33	5·2 ±0·25	7·1 ±0·29	8·1 ±0·45	9·8 ±0·47	10·6 ±0·75
placebo 1	5·1 ±0·43	4·5 ±0·41	7·8 ±0·30	9·3 ±0·52	11·2 ±0·56	12·2 ±1·00	11·8 ±0·90
placebo 2	5·5 ±0·91	5·6 ±1·29	7·9 ±0·86	10·1 ±0·84	11·7 ±0·96	12·1 ±0·96	13·5 ±0·52
			Hypertensives†				
atenolol	4·0 ±0·36	3·9 ±0·37	7·1 ±1·14	7·7 ±1·19	9·9 ±1·75	11·0 ±1·41	11·7 ±1·42
atenolol+ chlorthal.	4·0 ±0·33	3·8 ±0·37	6·3 ±1·01	7·3 ±0·92	9·1 ±1·53	9·2 ±0·82	9·0 ±0·98
placebo 1	4·4 ±0·42	4·3 ±0·38	7·5 ±0·78	7·5 ±1·31	9·7 ±0·98	10·5 ±0·99	12·1 ±1·56
placebo 2	4·9 ±0·53	4·5 ±0·24	8·2 ±1·14	8·6 ±0·84	9·6 ±0·93	12·1 ±1·48	14·5 ±0·72

*Age = 41 ± 2·1 years; height = 174 ± 2·3 cm; weight = 73 ± 4·2 kg. Mean arterial pressure (mmHg) atenolol = 88, atenolol + chlorthalidone = 88, placebo 1 = 90, placebo 2 = 96.
†Age = 38 ± 2·4 years; height = 176 ± 0·7 cm; weight = 89 ± 4·3 kg. Mean arterial pressure (mmHg) atenolol = 102, atenolol + chlorthalidone = 102, placebo 1 = 115, placebo 2 = 113.
Atenolol dose 50 mg 24 h before plus 50 mg 2 h prior to exercise; atenolol 50 mg + chlorthalidone 12·5 mg, same protocol. Study was fully randomized and double-blind.

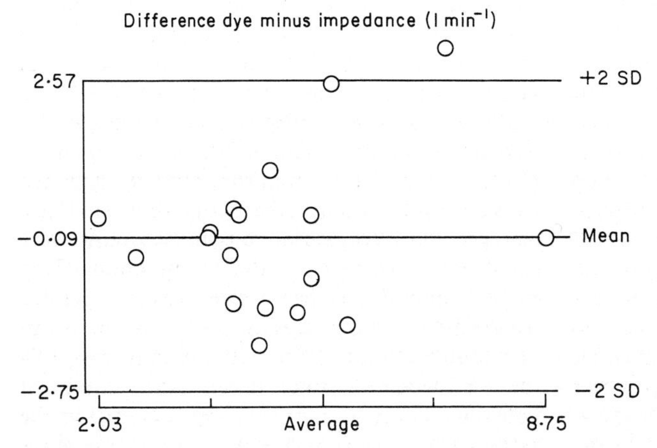

Figure 8 Comparison of dye vs impedance cardiac output in essential hypertensive patients.

but under exactly similar conditions. On the first morning, output averaged 5·1 ± 0·22 $l\,min^{-1}$ and on the second morning, 5·0 ± 0·26 $l\,min^{-1}$, the standard deviation of the differences being 10% of the arithmetic mean value. A plot according to Bland and Altman[19] revealed that, with a few exceptions, differences between the two measurements moved around a mean of only 0·05 $l\,min^{-1}$ (Fig. 9).

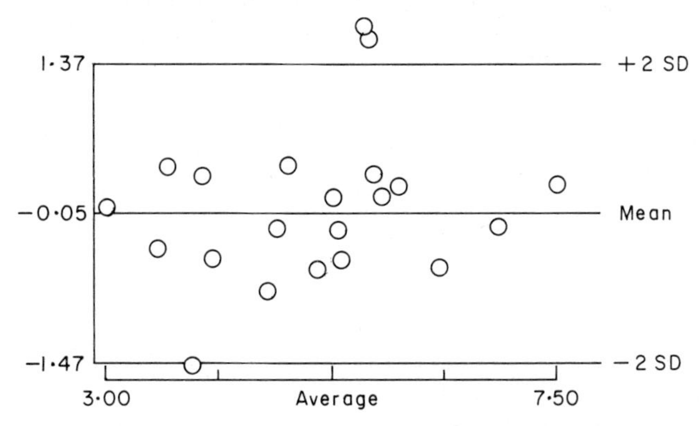

Figure 9 Comparison of two cardiac output measurements by impedance technique when data were obtained on 2 different days but under otherwise similar conditions.

These data are better than those provided by Stick and Büchsel[20] and underscore the potential of impedance cardiography as a highly reproducible method to assess serial haemodynamic changes in a hypertensive population.

There has been some debate about the haemodynamic mechanisms underlying the pathogenesis of early or borderline hypertension: both increased cardiac output and increased vascular resistance have been implicated[21]. In most studies, however, cardiac output was measured by invasive methods or in subjects who were aware of being hypertensive. We decided, therefore, to apply the non-invasive impedance technique in a 'naive' population of youngsters who did not know their blood pressure. From a survey of male students entering medical school, 30 subjects were selected after blood pressure had been measured on three different occasions. On the basis of their systolic pressure at the third examination they were divided into two groups, group I having a systolic pressure below 120 mmHg and group II having a systolic pressure above 140 mmHg but below 160 mmHg. In both groups, cardiac output was measured by the dye dilution technique and by impedance at the first examination, and by impedance only during a follow-up examination 2 years later. The results of this study are summarized in Table 6. Despite the difference in blood pressure, cardiac output was comparable in both groups; although dye dilution data could not be obtained from all subjects, the available results supported the same trend as was seen with impedance cardiography. After 2 years of follow-up, nearly identical values for cardiac output were registered as during the initial examination. Thus, using a non-invasive approach we were able to demonstrate that cardiac output plays no role in determining blood pressure in young people with only very mild elevation of (systolic) pressure.

Table 6 Mean blood pressure and cardiac output ($\pm$SEM) in two groups of young subjects with different levels of systolic pressure. Study 2 was carried out two years after study 1

	Systolic pressure (mmHg)	Diastolic pressure (mmHg)	Cardiac output-impedance (l min⁻¹).m⁻²
		Group I	
Study 1 (N=17)	116±2	72±1	3·2±0·2
Study 2 (N=11)	120±3	75±3	3·1±0·2
		Group II	
Study 1 (N=13)	139±4	79±2	3·1±0·1
Study 2 (N=10)	129±4	76±2	3·0±0·3

Subsequently, we evaluated whether impedance cardiography would yield meaningful results when hypertensive patients were exposed to a variety of stimuli expected to affect their blood pressure and which ranged from isometric exercise to certain types of treatment.

Firstly, in 34 patients the effects of sustained handgrip (25% of maximal strength for 3 min) were studied. The increase in blood pressure observed during the challenge was associated with a 34% increase in heart rate ($P<0.001$) and an 18% increment in cardiac output ($P<0.001$). Stroke volume, as determined by impedance measurements, did not change significantly. These data are in agreement with those of others who employed alternative methods to assess cardiac function during isometric exercise[22,23].

Next, we used impedance cardiography during tilt, a procedure which is usually associated with a variable decrease in blood pressure. To this end, seven patients with stable essential hypertension were tilted to 45° for 5 min. When we compared pre-tilt and post-tilt data, a decline in stroke volume ranging from 6 to 55% (mean 35%) was observed; cardiac output fell from 8 to 46% (mean 26%). Once again, these results agree well with those described in the literature both for impedance[24,25] and for invasive techniques[21,26].

Finally, we attempted to alter the haemodynamic situation by specific antihypertensive treatment in 49 patients with essential hypertension and, in addition to blood pressure measurements, we documented change in cardiac output by means of the impedance method. Only patients in whom blood pressure could not be controlled with one drug alone were eligible for the present analysis. Patients were divided into four groups: group A (N=17) received as initial treatment a direct vasodilator (hydralazine) to which a beta-blocker (propranolol) was added later. In group B (N = 12) the same drugs were used but in reverse order. Group C (N=12) first received the alpha-blocker, prazosin, and as a second drug the beta-blocker, atenolol. Group D (N=18) was initially treated with atenolol and received prazosin as the second drug. The results of this study are summarized in Table 7. There was a clear tendency for cardiac output to rise when hydralazine was given and to fall when propranolol was administered, irrespective of the order in which the drugs were prescribed. Changes in cardiac output appeared to be entirely due to concomitant changes in heart rate, stroke volume being unaffected. In the two groups of patients receiving atenolol and prazosin, cardiac output also was reduced by the beta-blocker and raised by prazosin. This time, however, the prazosin-induced rise in output was due to an increased stroke volume. Even though these patterns look good and confirm data from the literature[27–29], it should be noted that in at least six subjects we found surprisingly low values for cardiac output, ranging from 1·5 to 2·8 l min^{-1}. Nevertheless, percentage changes during treatment

Table 7 Mean arterial pressure (MAP) and cardiac output (CO) in four groups of essential hypertensive patients with different treatment modalities. All data given as mean ± SEM

	MAP (mmHg)	CO (l min^{-1})
Group A (N=17)		
Control	133±3	4·2±0·3
Hydralazine	117±3	4·8±0·5
Hydralazine + propranolol	104±2	3·6±0·3
Group B (N=12)		
Control	140±4	4·4±0·4
Propranolol	125±4	3·2±0·4
Propranolol + hydralazine	104±3	3·8±0·4
Group C (N=12)		
Control	134±4	4·3±0·2
Prazosin	122±4	4·9±0·4
Prazosin + atenolol	108±2	4·0±0·4
Group D (N=8)		
Control	146±8	6·2±0·2
Atenolol	121±7	4·6±0·2
Atenolol + prazosin	115±7	5·4±0·2

were comparable to those obtained in the other patients. Our findings thus indicate that impedance cardiography is suitable for monitoring sequential changes but that, in patients in this situation, some caution may be necessary in interpreting absolute data. When we combine the results of all studies in our hypertensive population we can conclude that the impedance technique has the advantage of providing highly reproducible results, which offers the possibility of following long-term haemodynamic changes non-invasively. Indeed, our data indicate that impedance-derived values for stroke volume and cardiac output show the expected directional shifts during various procedures that tend to affect the haemodynamic setting, e.g., a fall in cardiac output during tilt or during beta-blockade and a rise during vasodilator treatment. Despite these promising results there may be a problem with absolute data.

Other impedance cardiograph approaches in health and disease

These have been reviewed recently in terms of theoretical considerations[30,31] and technical extensions/computerization in the clinical environment[32]. Some controversy has arisen concerning the outcome of a theoretical approach different from the Kubicek model and proposed by Sramek[33,34]. The latter model assumes the thorax is shaped like a truncated cone rather than the cylinder of Kubicek, and L is not measured, but is derived from a nomogram based on sex, height and weight. Blood resistivity is also incorporated into the nomogram. A modified version of this theory[35] has been incorporated into a new instrument, NCCOM3 (BoMed Medical Manufacturing, Irvine, CA). Recent comparisons of the SV data in normal subjects obtained by the Kubicek model vs the Sramek model have indicated that the NCCOM3 system will overread[36,37], provide non-systematic results in hypertension of pregnancy[38], and give lower values at increasing levels of exercise[39] than the indirect Fick (CO_2 rebreathing) method. On the other hand, other laboratories consider the NCCOM3 will give estimates of cardiac output as satisfactory as other standard techniques used clinically under normal conditions and in critically ill patients[37,40].

Conclusions

We conclude from this review that the Kubicek model, as used in our laboratories, provides fair reproducibility for stroke volume and cardiac output measurement better than the 9–10% generally agreed necessary for decisions in clinical work. Carefully handled, and with due regard to its limitations in determining absolute values, it appears to be accurate enough for research into physiological mechanisms at rest and exercise in man and animals. This applies to neonatal man and very small animals, provided the cardiopulmonary system is free of shunts and cardiac valve abnormalities, where there is still evaluative work to be done[42]. It appears to be useful in uncomplicated hypertension, but will need further evaluation for use in complicated, long standing hypertension, severe cardiac failure, in pregnancy[38], and in intensive care wards when patients are being artificially ventilated[42].

We are grateful to Dr Howard Smith of ICI Pharmaceuticals (Australia) for support concerning the hypertension studies in man. The expert laboratory support of Ms Pam Glenfield, Ms Adele Buchanan and Clinical Associate Professor Ted Hennessy is also acknowledged. Mrs Ruth Barrett was a tireless and cheerful support with the manuscript.

References

[1] Vigouroux R. Sur la résistance électrique considérée comme signe clinique. Prog Med 1888; 16: 45.

[2] Cremer M. Uber die Registrierung mechanischer vorgange auf electrischen Wege, speziell mit hilfe des Saitengalvanometers und saiten Electrometers. Muenchen Med Wochenschr 1907; 33: 1629–30.

[3] Atzler E, Lehman G. Uber ein neues Verfahren zur Darstellung der Herztätigkeit. Arbeitsphysiologie 1932; 5: 636–80.

[4] Nyboer J. Plethysmography, impedance. In: Glasser O, ed. Medical physico, Vol III. Chicago: Yr Book Publ Inc 1960: 459–71.

[5] Moskulendo Y, Bayevskiy RM, Gazenko OG. Technique for investigating blood circulation in the brain under conditions of an altered gravitational field. In: Problems of space biology, Vol I. Moskva: Izd-Vo An SSSR 1962: 400–4.

[6] Kubicek WG, Karnegis JN, Patterson RP, Witsoe DA, Mattson RH. Development and evaluation of an impedance cardiac output system. Aerospace Med 1966; 37: 1208–12.

[7] Patterson RP. Cardiac output determinations using impedance plethysmography. MSc Thesis, University of Minnesota, Minneapolis, Minnesota, U.S.A., 1965.

[8] Quail AW, Traugott FM. Effects of changing haematocrit, ventricular rate and myocardial inotropy on the accuracy of impedance cardiography. Clin Exp Pharmacol Physiol 1981; 8: 335–43.

[9] Geddes LA, Sadler C. The specific resistance of blood at body temperature. Med Bio Eng Comput 1973; 11: 335–9.

[10] Mohapatra SN, Costeloe KL, Hill DW. Blood resistivity and its implications for the calculation of cardiac output by the thoracic electrical impedance technique. Intens Care Med 1977; 3: 63–7.

[11] Mohapatra SN, Hill DW. The changes in blood resistivity with hematocrit and temperature. Eur J Intens Care Med 1975; 1: 153–62.

[12] Denniston JC, Maher JT, Reeves JT, Cruz JC, Cymerman A, Grover RF. Measurements of cardiac output by electrical impedance at rest and during exercise. J Appl Physiol 1976; 40: 91–5.

[13] Keim HJ, Wallace JM, Thurston H, Case DB, Drayer JIM, Larah JH. Impedance cardiography for determination of stroke index. J Appl Physiol 1976; 41: 797–9.

[14] Quail AW, Traugott FM, Porges WL, White SW. Thoracic resistivity for stroke volume calculation in impedance cardiography. J Appl Physiol 1981; 50: 191–5.

[15] Traugott FM, Quail AW, White SW. Evaluation of blood resistivity in vivo for impedance cardiography in man, dog and rabbit. Med Biol Eng Comput 1981; 19: 547–52.

[16] Gotshall RW, Breay-Pilcher JC, Boelcskevy BD. Cardiac output in adult and neonatal rats utilizing impedance cardiography. Am J Physiol 1987; 253: H1298–304.

[17] Traugott FM, Quail AW, White SW, Letchford PJ, Moore PG. Impedance cardiography: clinical limitations and accuracy. Proceedings of the Vth International Conference on Electrical Bioimpedance, Tokyo 1981: 21–4.

[18] Wong DH, Onishi R, Tremper KK *et al.* Thoracic bioimpedance and Doppler cardiac output measurement: learning curve and interobserver reproducibility. Crit Care Med 1989; 17: 1194–8.

[19] Bland MJ, Altman DG. Statistical methods for assessing agreement between two methods of clinical measurement. Lancet 1986; i: 307–10.
[20] Stick C, Büchsel R. Impedance cardiography: the reproducibility of stroke volume measurement under conditions of mass examination. Basic Res Cardiol 1978; 73: 627–36.
[21] Birkenhäger WH, De Leeuw PW, Schalekamp MADH. Control mechanisms in essential hypertension. 2nd edit. Amsterdam: Elsevier Biomedical Press BV, 1982.
[22] Laird WP, Fixler DE, Huffines FD. Cardiovascular response to isometric exercise in normal adolescents. Circulation 1979; 59: 651–4.
[23] Peter CA, Jones RH. Effects of isometric handgrip and dynamic exercise on left-ventricular function. J Nucl Med 1980; 21: 1131–8.
[24] Smith JJ, Bush JE, Wiedmeier VT, Tristani FE. Application of impedance cardiography to study of postural stress. J Appl Physiol 1970; 29: 133–7.
[25] Thangarajah N, Hames T, Mubako H, Patel J, MacLennan WJ, The use of impedance cardiography in the young and elderly during postural stress. Age Ageing 1980; 9: 235–40.
[26] Lewis ML, Christianson LC. Behavior of the human pulmonary circulation during head-up tilt. J Appl Physiol 1978; 45: 249–54.
[27] Koch-Weser J. Hydralazine. N Engl J Med 1976; 295: 320–3.
[28] Holland OB, Kaplan NM. Propranolol in the treatment of hypertension. N Engl J Med 1976; 294: 930–6.
[29] Lund-Johansen P. Haemodynamic long-term effects of prazosin plus tolamolol in essential hypertension. Br J Clin Pharmacol 1977; 4: 141–5.
[30] Lamberts R, Visser KR, Zijlstra WG. Impedance cardiography. Assen, The Netherlands: Van Gorcum, 1984.
[31] Penny BC. Theory and cardiac applications of electrical impedance measurements. CRC Crit Rev Biomed Eng 1985; 13: 227–81.
[32] Buell JC. A practical, cost-effective, non-invasive system for cardiac output and hemodynamic analysis. Am Heart J 1988; 116: 657–64.
[33] Sramek BB, Rose DM, Miyamato A. Stroke volume equation with a linear base impedance model and its accuracy as compared to thermodilution and magnetic flowmeter techniques in humans and animals. In: Proceedings of the Sixth International Conference on Electrical Bioimpedance. Zadar, Yugoslavia, 1983: 38.
[34] Sramek BB. Electrical bioimpedance. Med Elect 1983; 23: 95–105.
[35] Bernstein DP. A new stroke volume equation for thoracic bioimpedance: theory and rationale. Crit Care Med 1986; 14: 904.
[36] de May C, Enterling D. Noninvasive assessment of cardiac performance by impedance cardiography: disagreement between two equations to estimate stroke volume. Aviat Space Environ Med 1988; 59: 57–62.
[37] Gotshall RW, Wood VC, Miles DS. Comparison of two impedance cardiographic techniques for measuring cardiac output in critically ill patients. Crit Care Med 1989; 17: 806–11.
[38] Easterling TR, Benedetti TJ, Carlson KL, Watts DH. Measurement of cardiac output in pregnancy by thermodilution and impedance techniques. Br J Obstet Gyn 1989; 96: 67–9.
[39] Smith GA, Russell AE, West MJ, Chalmers JP. Automated non-invasive measurement of cardiac output: comparison of electrical bioimpedance and carbon dioxide rebreathing techniques. Br Heart J 1988; 59: 292–8.
[40] Salandin V, Zussa C, Risica G *et al.* Comparison of cardiac output estimation by thoracic electrical bioimpedance, thermodilution, and Fick methods. Crit Care Med 1988; 16: 1157–8.
[41] Miles DS, Gotshall RW, Golden JC, Tuuri DT, Beekman RH, Dillon T. Accuracy of electrical impedance cardiography for measuring cardiac output in children with congenital heart disease. Am J Cardiol 1988; 61: 612–6.
[42] Preiser JC, Daper A, Parquier J-N, Contempré B, Vincent J-L. Transthoracic electrical bioimpedance versus thermodilution technique for cardiac output measurement during mechanical ventilation. Intens Care Med 1989; 15: 221–3.

Systolic time intervals

H. BOUDOULAS

Professor of Medicine and Pharmacy, Division of Cardiology The Ohio State University, Columbus, Ohio, U.S.A.

KEY WORDS: Systolic time intervals, ventricular function, cardioactive drugs, ventricular filling, coronary flow.

The systolic time intervals (STI) offer temporal description of the sequential phases of cardiac cycle which are influenced physiologically by the same variables as affect other measures of left ventricular (LV) performance. The STI hence offer a measure of ventricular function which augments other measures of ventricular performance. Because of the extreme sensitivity of this variable and the ease of its measurement the STI are well suited for studying the effect of pharmacologic agents upon the heart.

Introduction

Left or right ventricular (LV or RV) contraction results in force generation, shortening of the myocardial walls, and ejection of blood into the aorta or pulmonary artery. Further, LV or RV contraction has a defined temporal relationship[1]. Contractile function abnormalities of the LV or RV may result in diminished force generated by the myocardium, diminished degree of myocardial wall shortening, diminished volume of blood delivered to the circulation, and abnormal temporal relationships of the phases of contraction.

Indices of ventricular performance are based on the capacity of the ventricle to pump blood (stroke volume, cardiac output), the ability to generate force (pressure, maximal dP/dt), the ability to shorten with each contraction (ejection fraction), the temporal relationships of contraction (systolic time intervals), and combinations of these variables (Fig. 1).

Systolic time intervals

METHODOLOGY

Most tests of ventricular performance deal with stroke volume-cardiac output, force and/or distance. The systolic time intervals (STI) are the only tests for which the sole measure of ventricular function is time.

The STI, as originally defined, are determined from simultaneous high-speed recordings of the electrocardiogram, the phonocardiogram, and the carotid arterial pulse tracing. Three primary measurements of the phases of systole are determined. Electromechanical systole (QS_2) spans the period from the onset of the QRS complex on the electrocardiogram to the closure of the aortic valve as denoted in the second heart sound (Fig. 2). The left ventricular ejection time (LVET) is the phase of systole during which the left ventricle ejects blood into the arterial system (Figs 1, 2). The pre-ejection period (PEP) is the interval from the onset of ventricular depolarization to the beginning of ejection (Figs 1, 3); PEP is derived by subtracting the LVET from the QS_2 interval[3,4] (Fig. 2). Calculation of PEP in this manner discounts the error resulting from the delay in transmission of the arterial pulse from the proximal aorta to the point of its detection over the carotid artery. In practice, the standard lead which elicits the earliest onset of electrical depolarization is selected. The phonocardiogram should be recorded over the upper precordium with the frequency range 100–500 Hz. The carotid arterial pulse must be recorded with high-fidelity instrumentation, preferably a strain gauge transducer with a frequency of 0·1 to 30 Hz and at least a 2 s time constant.

For the best definition of STI, one must be certain that the recordings delineate a sharp inscription of the initial high-frequency vibrations of the aortic component of the second heart sound on the phonocardiogram, and a clearly discernible upstroke and notch on the carotid arterial pulse tracing. A minimum paper speed of 100 mm s^{-1} is necessary for accurate determination of the STI. Calculations of STI are derived from the mean of at least 10 consecutive beats obtained while the patient

Address for correspondence: Harisios Boudoulas, M.D., The Ohio State University, Division of Cardiology, 647 Means Hall, 1664 Upham Drive, Columbus, Ohio 43210, U.S.A.

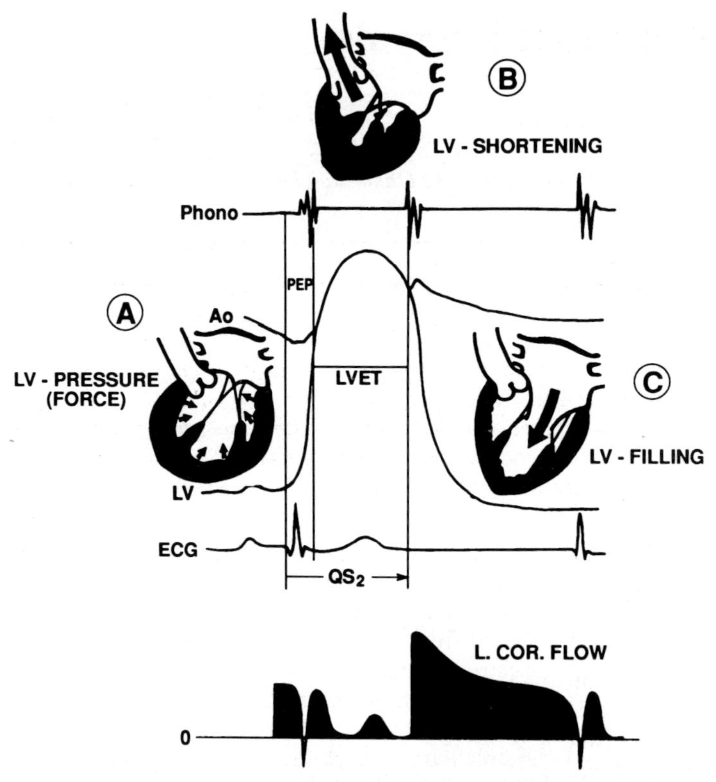

Figure 1 Simultaneous recordings of the phonocardiogram (phono), aortic (Ao) pressure, left ventricular (LV) pressure, electrocardiogram (ECG) and left coronary artery flow (L Cor. Flow) — schematic presentation. The LV at different phases of the cardiac cycle is also shown. The LV during isovolumic contraction time (A) generates force; during the ejection phase (B) shortens its walls and ejects blood into the aorta. Further, LV contraction has a defined temporal relationship. The time interval during which the LV generates force is provided by the pre-ejection period (PEP), and the time during which the LV ejects blood is provided by the left ventricular ejection time (LVET).

Indices of LV performance are based on the capacity of the LV to pump blood into the aorta (stroke volume — cardiac output), the ability to generate force (pressure, dP/dt), the ability to shorten with each contraction (ejection fraction), the temporal relationship of contraction (systolic time intervals), and combinations of these parameters.

Diastolic time begins with the end of systole (C); diastolic time is an important factor which determines myocardial blood flow and LV ventricular filling. QS_2 = total systolic time.

is supine during quiet respiration. The validation of the technical and physiologic principles underlying the measurement of STI have been summarized in previous publications from our own and other laboratories[3–19].

STI also can be determined from simultaneous recordings of the M-mode echocardiogram of the aortic valve and the electrocardiogram employing the measurement of the interval from the onset of ventricular depolarization to the opening of the aortic valve (PEP), and that between the opening and closure of the aortic valve (LVET)[20,21]. QS_2 can be calculated by adding PEP to LVET. The echocardiographic visualization of the opening and closing of the aortic valve, however, is present in only 80–85% of adult patients (Fig. 2).

The Doppler echocardiographic technique can also be used to calculate STI. PEP is the interval from the beginning of the QRS to the beginning of the aortic flow, and LVET is the interval from the

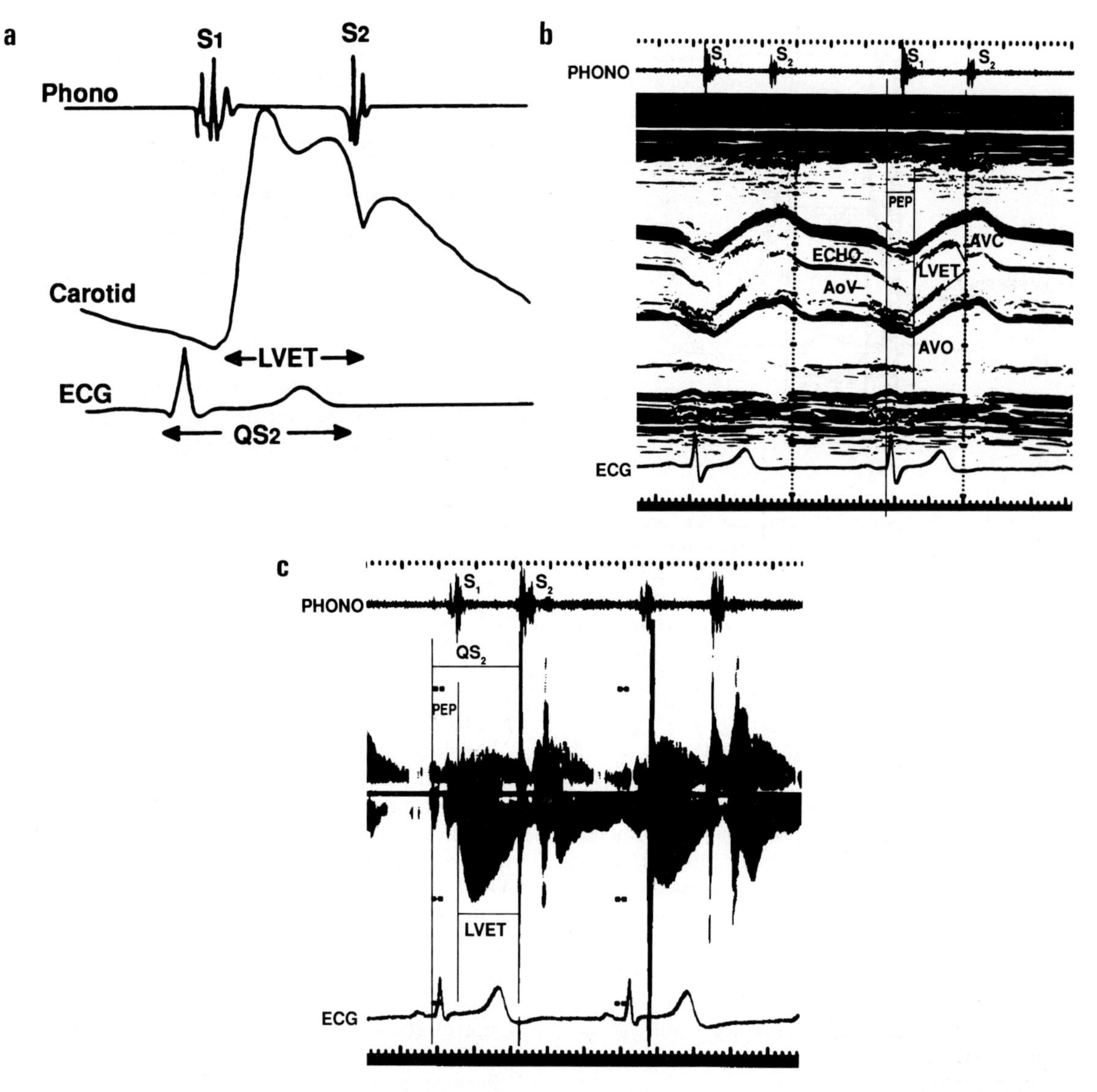

Figure 2 (a) Measurements of the systolic time intervals (STI) from simultaneous recordings of the electrocardiogram (ECG), phonocardiogram (phono), and carotid arterial pulse tracing (carotid). PEP = $QS_2 - LVET$. (b) Determination of STI from the aortic valve echocardiogram (ECHO AoV) and the ECG. The aortic valve opening (AVO) and closure (AVC) are shown. (c) Determination of the STI from aortic flow using Doppler echocardiography. The PEP, LVET and QS_2 are shown, S_1, S_2 = first and second heart sounds respectively.

beginning to the termination of the aortic flow (Fig. 2). QS_2 can be calculated by adding PEP to LVET. Using Doppler techniques, aortic flow and pulmonic flow can be easily obtained and thus, LV and RV STI can be calculated.

The three approaches are complementary, so that when the temporal landmarks for the STI are not definable by one method, or a method is not available, the other can be used as an alternative approach. It must be emphasized that precise definition of the aortic valve opening and closure and the beginning and termination of the aortic blood flow is mandatory when echocardiography and Doppler echocardiography are used for STI measurements: a paper speed of 100 mm s^{-1} is also necessary for accurate determination of STI.

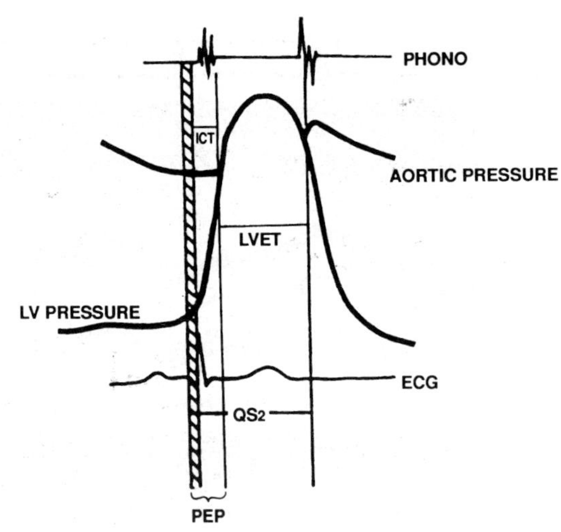

Figure 3 Schematic presentation of the phonocardiogram (phono), aortic pressure, left ventricular (LV) pressure. The pre-ejection period (PEP) — time from the beginning of the QRS to the beginning of the increase in aortic pressure — is equal to the electromechanical delay (hatched area) plus the isovolumic contraction time (ICT). LV ejection time (LVET) is also shown.

Correction for heart rate

It has long been appreciated that all the components of STI vary inversely with heart rate. Thus, in order to interpret changes in the STI, correction must be made for variations in resting heart rate; the normal values for the resultant indices (PEPI, LVETI, QS_2I) are shown in Table 1[3]. Some investigators have suggested that changes of PEP with heart rate are minimal and thus correction for heart rate is not necessary[22,23]. Although these changes are small, they are statistically significant and correction of the PEP for heart rate increases the sensitivity of the method especially in clinical pharmacology.

When atrial fibrillation is present, studies have shown that only beats with R–R intervals greater than 800 ms (heart rate $\leqslant 75$ beats min^{-1}) should be averaged[24].

Assessment of LV performance

The STI offer temporal description of the sequential phases of the cardiac cycle which are influenced physiologically by the same variables as affect other measures of global LV performance. STI hence offer a measure of ventricular function which augments other measures of LV performance, and which amplifies our understanding of the altered contractile events which accompany LV decompensation[1,3,4,6–8,10,11,16–19]. Several clinical and experimental studies have shown that interventions which alter LV performance produce directionally similar changes in the STI and other non-invasive and invasive indices of LV performance[1,3,4,6–8,10,11,16–19]. *Thus, STI can be used to detect directional changes of LV performance.*

When LV dysfunction occurs, regardless of the cause, PEPI lengthens, LVETI shortens, QS_2I remains relatively constant and PEP/LVET increases[3,4].

The prolongation of PEPI can be attributed almost entirely to a diminished rate of increase in LV pressure during isovolumetric contraction[3,4,6–8,11,16,25]. The diminished rate of LV pressure rise (LV dp/dt) results in a prolongation of the isovolumic contraction time and PEP (Fig. 3). This occurs because PEP is equal to electromechanical delay (interval from the beginning of the QRS to the beginning of the ventricular contraction) plus the isovolumic contraction time. The electromechanical delay is constant within a wide range of heart rates. Thus, changes in PEP are largely dependent on the isovolumic contraction time; interventions that prolong the isovolumic contraction time will prolong PEP and vice-versa.

The shortening of LVETI with LV dysfunction is more complex. A primary factor must be the relative lengthening of PEP, which induces a delayed onset of ejection. During ejection, the velocity of myocardial fibre-shortening (Vcf) is diminished when LV failure is present, a condition that theoretically should prolong LVETI. However, the extent of fibre shortening (ejection fraction) is also reduced and this tends to shorten the LVETI. Further, LVETI is related to stroke volume; decrease in stroke volume will tend to shorten the LVETI and vice-versa[3,4,26–29].

Because in LV dysfunction PEP is lengthened and LVET is shortened, the ratio of PEP/LVET provides a more useful index of overall LV performance[3,4] (Fig. 4).

CLINICAL APPLICATIONS

Myocardial disease

Numerous studies of patients with myocardial disease have shown abnormal STI in patients with angiographically abnormal ejection fraction or abnormalities of other measures of LV performance[3,4,6–8,10,11,13,16–19]. In these patients, STI can define the severity of LV muscle dysfunction both as a baseline for initial assessment and for long-term follow-up or evaluation of therapy.

Table 1 *Calculation of systolic time interval index (I)* values from resting regression equations*

Sex	Equation	Mean normal values (ms) ± 1 SD	Range
Male	$QS_2I = 2{\cdot}1\ HR + QS_2$	546 ± 14	518–574
Female	$QS_2I = 2{\cdot}0\ HR + QS_2$	459 ± 14	521–577
Male	LVETI = 1·7 HR + LVET	459 ± 10	393–433
Female	LVETI = 1·6 HR + LVET	418 ± 10	393–438
Male	PEPI = 0·4 HR + PEP	131 ± 10	111–151
Female	PEPI = 0·4 HR + PEP	133 ± 10	113–153

HR = Heart rate, QS_2I, LVETI, PEPI = total electromechanical systole, left ventricular ejection time, and pre-ejection period, respectively, corrected for heart rate.
*From Weissler *et al.* Circulation 1968; 37: 149–59[3].

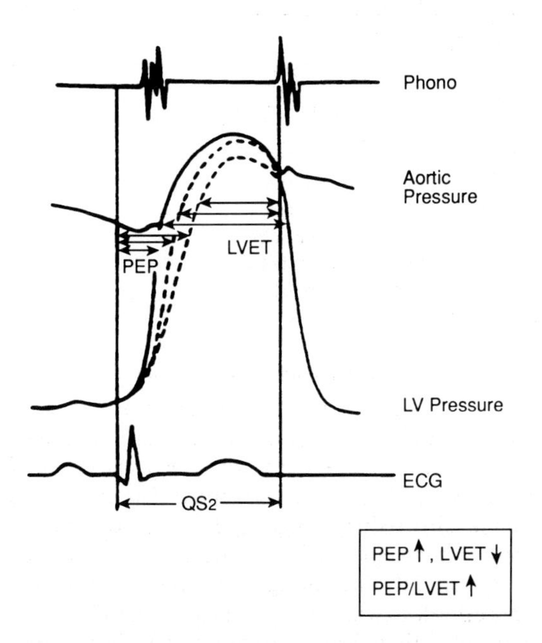

Figure 4 Relationship of systolic time intervals (STI) to the cardiac cycle (normal, solid line). The dotted lines indicate the changes induced in left ventricular (LV) dysfunction where the pre-ejection period (PEP) is prolonged and left ventricular ejection time (LVET) is shortened. Due to PEP prolongation and LVET shortening the PEP/LVET is also increased.

Coronary artery disease

STI have also been used to estimate prognosis both in patients with angina pectoris and in patients who have recovered from myocardial infarction[24–33]. It is known that the status of LV function is a major prognostic indicator in patients with coronary artery disease. STI can stratify patients at low and high risk based on the PEP/LVET abnormality. Patients with an abnormal PEP/LVET have a significantly worse 5-year prognosis than patients with normal PEP/LVET. In patients with coronary artery disease therapeutic interventions usually have improved survival only in high risk patients. Indeed, studies from our laboratory have shown that therapy with beta-blocking drugs or coronary bypass surgery improve survival only in patients with abnormal PEP/LVET[34,35]. These results are in agreement with more extensive randomized double-blind prognostic studies.

Arterial hypertension

It is generally appreciated that the progressive increase in LV mass that accompanies chronic arterial hypertensive disease results in LV dysfunction and, ultimately, congestive heart failure[36–38]. In a study from our laboratory, the relation of LV functional abnormalities to the extent of increase in LV mass has been evaluated[37]. Ninety patients with chronic systemic hypertension were compared with 41 normal subjects as determined by cardiac catheterization and angiography. LV mass was estimated from the M-mode echocardiogram. Patients were separated into three groups: those with LV mass of less than 2, (group I, n = 58), 2–4 (group II, n = 21) and more than 4 (group III, n = 11) standard deviations above mean normal. The ratio of PEP/LVET, percent shortening of the echocardiographic internal diameter (%ΔD) and

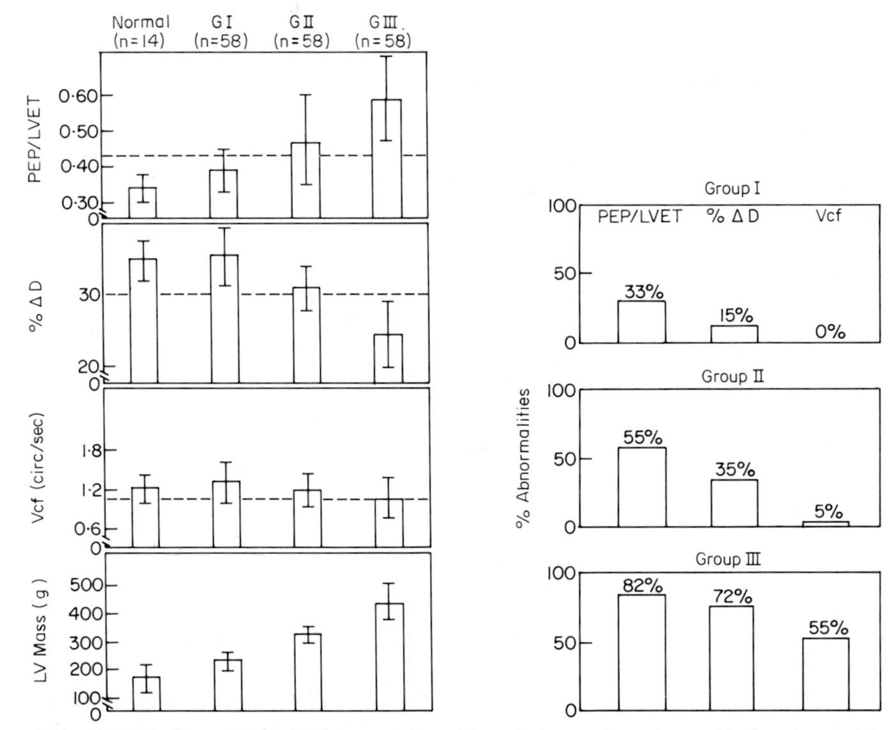

Figure 5 (a) Left ventricular (LV) mass and systolic performance in patients with chronic arterial hypertension PEP/LVET = pre-ejection period over left ventricular ejection time; %ΔD = percent shortening of the echocardiographic internal diameter; Vcf = mean rate of circumferential fibre shortening. Dashed lines show the upper normal limits for PEP/LVET and lower normal limits for %ΔD and Vcf. (b) Frequency of abnormalities, expressed as percentage of patients with abnormal measures in the three groups of patients. NS = non-significant. (Modified from[37]).

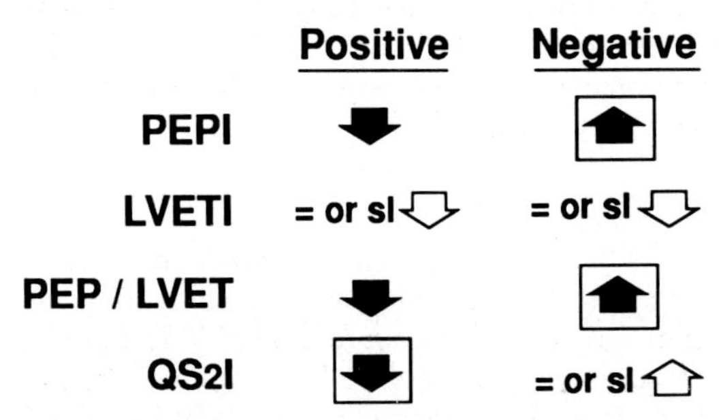

Figure 6 Changes of the systolic time intervals (STI) with positive and negative inotropic effect. PEP = pre-ejection period, LVET = left ventricular ejection time, QS_2 = total electromechanical systole. Arrows in the boxes indicate most significant effect.

velocity of circumferential myocardial fibre shortening (Vcf) were used as indexes of LV systolic performance. The frequency of abnormality, expressed as percentage of patients in groups I, II and III, was 33%, 55% and 85% for PEP/LVET, 15%, 35% and 72% for %ΔD, and 0%, 15% and 55% for Vcf. For each group, PEP/LVET was the most frequent abnormal measurement while Vcf was the least frequent abnormality (Fig. 5). Calculation of peak and end-systolic wall stress was used as an index of the adequacy of LV hypertrophy. The index was significantly reduced in group I, did not differ from control in group II and was significantly increased in group III, indicating that hypertrophy was appropriate to wall tension in groups I and II. It was concluded that the occurrence of LV dysfunction with increasing LV mass in patients with mild and moderate LV hypertrophy (groups I and II) reflects a deficiency in intrinsic contractile performance of the hypertrophied myocardium which can be detected with the STI. With a marked increase in LV mass (group III), inadequacy in LV hypertrophy relative to wall tension may also contribute to LV dysfunction.

CLINICAL PHARMACOLOGY

Pharmacologic agents with positive or negative inotropic effect produce significant changes in the

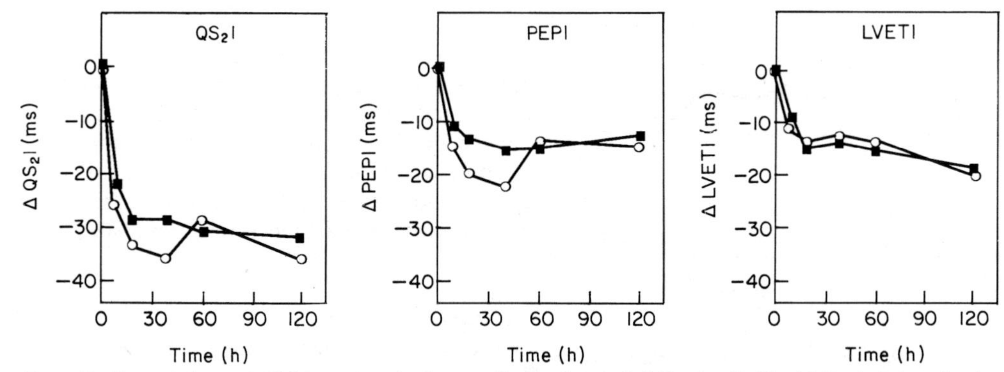

Figure 7 Temporal course of the response in the systolic time intervals following Cedilanid-D administration in 10 normal subjects and 11 patients with congestive heart failure. ΔQS_2I, ΔPEPI and ΔLVETI = changes of the total electromechanical systole, pre-ejection period and left ventricular ejection time corrected for heart rate respectively. (Modified from[42]).

STI. Positive inotropic drugs which shorten the isovolumic contraction time will shorten the PEP, and negative inotropic drugs which prolong the isovolumic contraction time will prolong the PEP[3,4,6–8,10,11,15,25,39,40].

Unlike PEPI, which responds in an opposite manner to positive and negative inotropic agents, LVETI is generally shortened by both. As mentioned previously, LVETI depends on the velocity of myocardial fibre shortening, the extent of fibre shortening and the stroke volume. The effect of the pharmacologic agents on the LVETI therefore, will depend on the effect of the drugs on the above factors[3,4]. Generally, with positive inotropic drugs the LVETI shortens slightly or remains unchanged; lengthening, however, can occur when a marked haemodynamic change occurs (that is, a large increase in the stroke volume). With negative inotropic drugs LVETI usually shortens.

The duration of the systole (QS_2I) is one of the relative constants in the circulatory system. Although most types of heart disease may produce profound alterations in cardiac performance manifested by directionally opposite changes in the PEPI, and LVETI, the QS_2I remains unchanged from normal unless high catecholamine levels or drug effects are present. *Because positive inotropic agents generally shorten both LVETI and PEPI, QS_2I is the most sensitive of the systolic time intervals in judging the presence of positive inotropic stimulation*[1,4]. A short QS_2 interval in the absence of pharmacologic agents suggests high intrinsic adrenergic tone.

As a general rule, the shortening of QS_2I is used to evaluate a positive inotropic effect and prolongation of PEPI or increase of PEP/LVET is used to evaluate a negative inotropic effect[3,4] (Fig. 6). *Thus, STI can give information for the overall LV performance (PEP/LVET) and for the presence of positive inotropic effect on the myocardium (QS_2I).*

Because changes in STI are easily discernible and can be measured non-invasively and repetitively, STI can be used to determine the rate and extent of the cardiac response to inotropic agents by serial observations. *Thus, STI are well suited for studying effects of pharmacologic agents upon the heart.*

Positive inotropic drugs

The clinical pharmacology of the various digitalis glycosides has been evaluated with the STI[41,42] (Fig. 7). These studies have shown that QS_2I is the most sensitive of the STI in assessing the presence of positive inotropic stimulation, whereas the haemodynamic response to the agent is reflected by the ratio of PEP/LVET.

The adrenergic subset of positive inotropic agents has also been studied with STI[43–48]. Furthermore, the STI have greatly enhanced the understanding of adrenergic receptor modulation, having shown rebound hypersensitivity to β-agonists following abrupt propranolol withdrawal in normal subjects[47,48] (Fig. 8).

Negative inotropic drugs

The effect of negative inotropic agents upon the STI has also been studied. In general, it appears that such agents induce the typical pattern of prolongation of PEPI and increase in PEP/LVET[1,4].

The effect of administration of lidocaine on LV performance was studied in our laboratory[49]. The greatest response in STI occurred at 3 min after

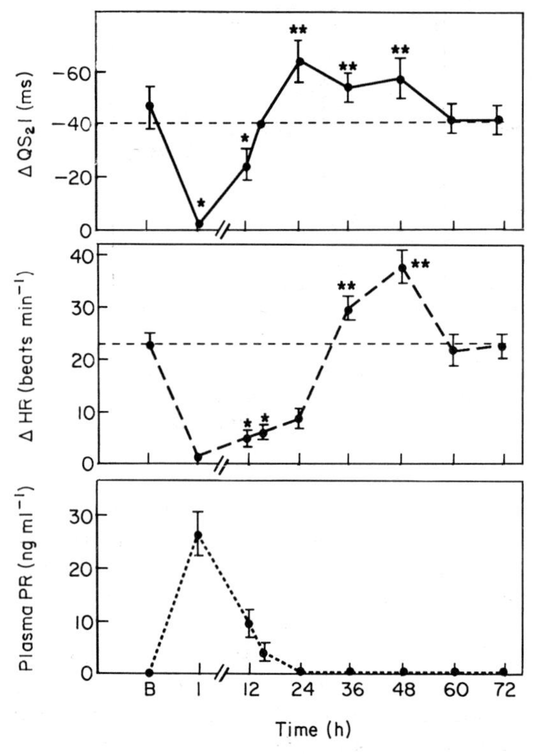

Figure 8 Percent change in total electromechanical systole corrected for heart rate (QS_2I) and heart rate (HR) due to isoproterenol administration (2 μg min^{-1} for 10 min) before (B), and up to 72 h after the last dose of propranolol (160 mg daily for 2 days) in six normal volunteers. The response to isoproterenol was significantly less than baseline ($* = P < 0{\cdot}05$) early after propranolol withdrawal, became significantly greater ($**P < 0{\cdot}05$) later after propranolol withdrawal, and finally returned to baseline. The time course of the inotropic blockade and hypersensitivity (QS_2I) was earlier than the chronotropic blockade (HR). Propranolol (PR) plasma concentration is also shown. (Modified from[48]).

intravenous injection of lidocaine, with values returning to baseline at 10 to 15 min. Administration of lidocaine produced a significant prolongation of PEPI and an increase of PEP/LVET. The effect of administering lidocaine after intravenous injection of propranolol was also studied. Although propranolol therapy alone increased the PEP/LVET, a further significant increase followed subsequent injection of lidocaine (Fig. 9).

It is known that disopyramide has a strong negative inotropic effect. This effect was manifested with dramatic changes in the STI[50] (Fig. 10). The dose-related increase in PEPI and PEP/LVET was found after disopyramide administration; the PEPI and PEP/LVET changes were directly related to the free but not to total disopyramide plasma concentrations[50]. Further, it was demonstrated that the negative inotropic effect of disopyramide is stereoselective; the R(−) enantiomer has greater negative inotropic effect than the S(+) enantiomer[51]. The negative inotropic effect of other antiarrhythmic drugs has also been demonstrated using STI[39].

Afterload-Preload

Other factors besides inotropy may alter STI. Thus, caution must be exercised to ensure that pharmacologic effects on the peripheral circulation do not confuse the interpretation of the STI. This is particularly important when marked changes in afterload and preload occur. Studies from our laboratory have shown that acute reduction in afterload with vasodilators in patients with congestive heart failure produced an increase in cardiac output and stroke volume and an improvement of LV performance as measured from STI. The correlation coefficient for vasodilator-induced changes in cardiac output vs changes in PEP/LVET was $-0{\cdot}82$ and for changes in stroke volume vs changes in PEP/LVET was $-0{\cdot}75$[52]. A controlled trial of the converting enzyme inhibitor enalapril in clinical cardiac failure showed, with treatment, a significant reduction in PEP/LVET parallel with improvement in exercise capacity and diminution of symptoms[53].

Acute increase in afterload prolongs the PEPI, has no effect or slightly prolongs the LVETI, and increases the PEP/LVET. The effect of chronic increase in afterload on the STI depends on the degree and adequacy of LV hypertrophy. Increased afterload without adequate LV hypertrophy will result in a prolongation of PEPI, in a shortening of the LVETI, and in an increase of the PEP/LVET. (Also see arterial hypertension). Alterations in preload produce significant changes in the STI. Increase in preload will shorten PEPI, prolong LVETI and decrease PEP/LVET; decrease in preload will produce opposite changes in PEPI, LVETI and PEP/LVET.

OTHER USES OF STI

STI can be used to study changes in LV performance occurring during haemodialysis[54]. STI are suitable for studying the effect of arteriovenous fistula on LV performance in patients undergoing haemodialysis. Studies also have shown that changes in preload occurring with upright posture

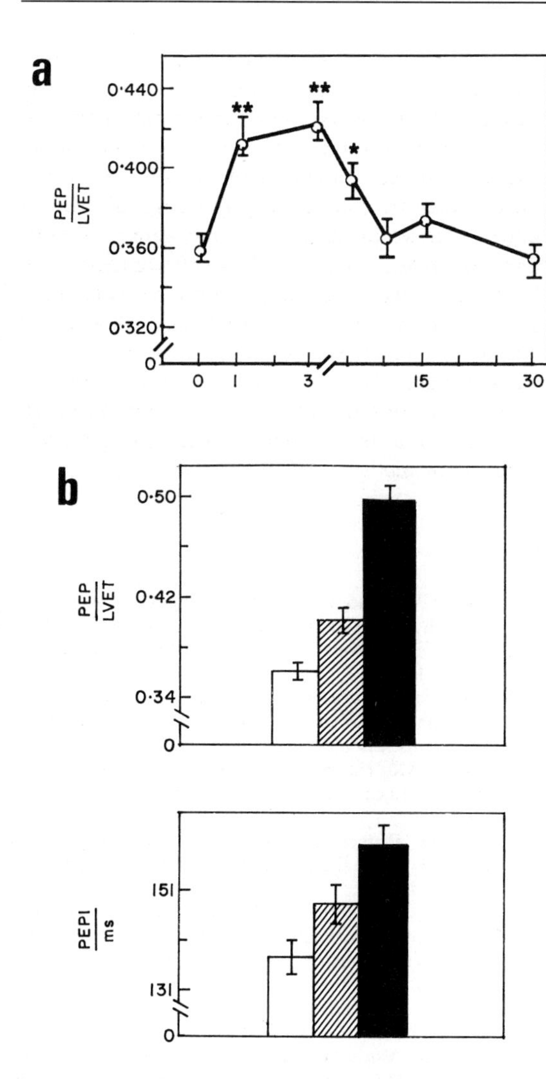

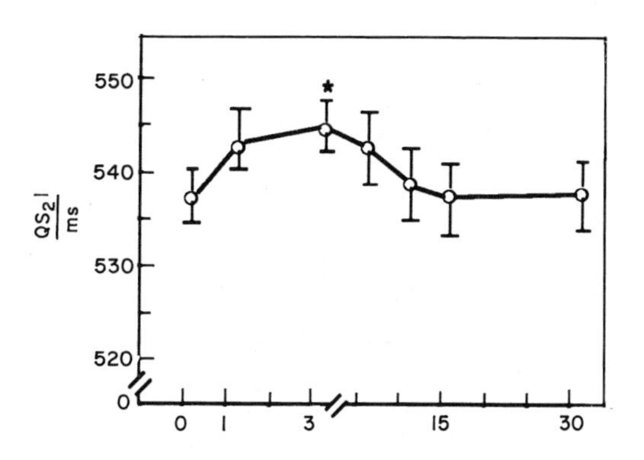

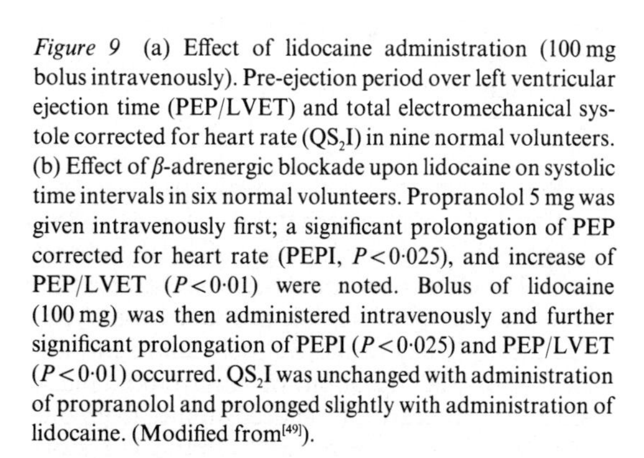

Figure 9 (a) Effect of lidocaine administration (100 mg bolus intravenously). Pre-ejection period over left ventricular ejection time (PEP/LVET) and total electromechanical systole corrected for heart rate (QS_2I) in nine normal volunteers. (b) Effect of β-adrenergic blockade upon lidocaine on systolic time intervals in six normal volunteers. Propranolol 5 mg was given intravenously first; a significant prolongation of PEP corrected for heart rate (PEPI, $P<0{\cdot}025$), and increase of PEP/LVET ($P<0{\cdot}01$) were noted. Bolus of lidocaine (100 mg) was then administered intravenously and further significant prolongation of PEPI ($P<0{\cdot}025$) and PEP/LVET ($P<0{\cdot}01$) occurred. QS_2I was unchanged with administration of propranolol and prolonged slightly with administration of lidocaine. (Modified from[49]).

produce changes in the PEP/LVET consistent with decreased cardiac output, and in the QS_2I consistent with positive inotropic effect[55–57].

Doppler techniques can be used to measure RV and LV STI. Measurements of the RV and LV STI may aid understanding of RV-LV interactions and to follow the natural history of LV and RV function in disease states[58–60]. The effect of pharmacologic agents (similar or dissimilar) and/or other interventions on RV and LV function in health and disease states can be studied and may enhance understanding of the dynamics of the systemic and pulmonary circulation.

Other studies have shown that in patients with hypertrophic cardiomyopathy or valvular aortic stenosis the LVETI can define patients with and

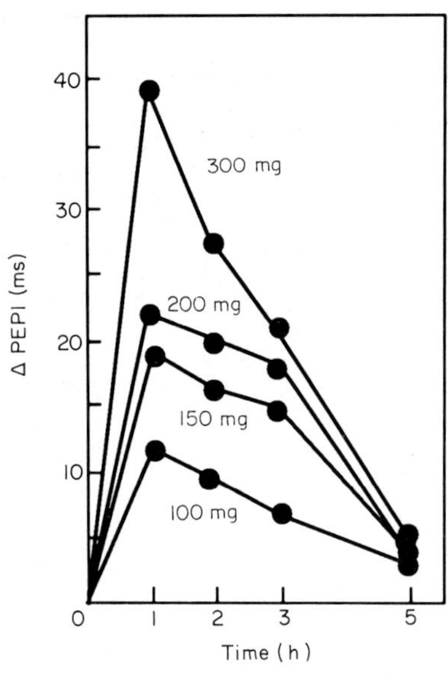

Figure 10 Dose related increased in pre-ejection period corrected for heart rate (ΔPEPI) after disopyramide administration in normal volunteers. (From[50]).

without significant gradient across the left ventricular outflow tract[4,61].

Prolongation of electrical systole (QT) is associated with a high incidence of ventricular arrhythmias and sudden death in patients with and without coronary artery disease. Traditionally, the heart rate is employed for correcting the duration of electrical systole and hence for identifying patients with prologation of the QT interval. Studies from our laboratory have demonstrated that the duration of the QT interval in normal subjects is consistently shorter and closely parallels the duration of electromechanical systole (QS_2) throughout the normal range of resting heart rate[62]. The close association between the duration of the QT and QS_2 intervals prompted the hypothesis that QT may be more closely linked physiologically to the duration of electromechanical systole than to heart rate. Indeed studies from our laboratory have shown that electrical systole greater than electromechanical systole ($QT > QS_2$) clinically defines a group of patients with prior myocardial infarction at high risk of death. The QT interval corrected for QS_2 was superior to the QT corrected for heart rate as a prognostic indicator[30].

Concluding remarks

In conclusion, indices of ventricular performance are based on the capacity of the ventricle to pump blood, the ability to generate force, the ability to shorten with each contraction, the temporal relationships of contractions, and combination of these variables (Fig. 1). STI offer a temporal description of the sequential phases of the cardiac cycle which are influenced physiologically by the same variables as affect other measures of global LV performance. Thus, STI reflect a dimension of LV function which is unique compared with other measures of ventricular performance. Because of the extreme sensitivity of the test and the ease of measurement, STI are ideally suited for studying the effects of pharmacologic agents upon the heart. *Indeed, this may represent the most useful application of the technique.*

References

[1] Binkley PF, Boudoulas H. Measurements of myocardial inotrophy. In: Cardiotonic drugs. A clinical survey. Leier CV, Ed. New York, Basel: Marcel Dekker Inc. 1986: 5–48.
[2] Boudoulas H, Rittgers SE, Lewis RP, Leier CV, Weissler AM. Changes in diastolic time with various pharmacologic agents. Circulation 1979; 60: 164–9.
[3] Weissler AM, Harris WS, Schoenfeld CD. Systolic time intervals in heart failure in man. Circulation 1968; 37: 149–59.
[4] Lewis RP, Rittgers SE, Forester WF, Boudoulas H. A critical review of the systolic time intervals. Circulation 1977; 56: 146–58.
[5] List WF, Gravenstein JS, Spodick DH. Systolic time intervals. International Boehringer Mannheim Symposium, Graz, Austria, Berlin, Heidelberg, New York: Springer-Verlag, 1980.
[6] Weissler AM. Interpreting systolic time intervals in man. J Am Coll Cardiol, 1983; 2: 1019–20.
[7] Talley RC, Meyer JF, McNay JL. Evaluation of the pre-ejection period as an estimate of myocardial contractility in dogs. Am J Cardiol 1971; 27: 384–91.
[8] Ahmed SS, Levinson GE, Schwartz CJ, Ettinger PO. Systolic time intervals as measures of the contractile state of the left ventricular myocardium in man. Circulation 1972; 46: 559–71.
[9] Van De Werf F, Piessens J, Kesteloot H, De Geest H. A comparison of systolic time intervals derived from the central aortic pressure and from the external carotid pulse tracing. Circulation 1975; 51: 310–6.
[10] Leier CV, Desch CE, Magorien RD *et al.* Positive inotropic effects of hydralazine in human subjects: comparison with prazosin in the setting of congestive heart failure. Am J Cardiol 1980; 46: 1039–44.
[11] Nakamura Y, Wiegner AW, Gaasch WH, Bing OHL. Systolic time intervals: assessment by isolated cardiac muscle studies. J Am Coll Cardiol 1983; 2: 973–8.
[12] Ulmer HE, Huepel EW, Weckesser G. Mechanocardiographic assessment of systolic time intervals in normal children. Bas Res Cardiol 1982; 77: 197–212.
[13] Stack RS, Sohn YH, Weissler AM. Accuracy of systolic time intervals in detecting abnormal left ventricular

performance in coronary artery disease. Am J Cardiol 1981; 47: 602–9.

[14] Levi GF, Ratti S, Cardone G, Basagni M. On the reliability of systolic time intervals. Cardiology 1982; 69: 157–65.

[15] Kupari M. Reproducibility of the systolic time intervals: Effect of the temporal range of measurements. Cardiovasc Res 1983; 17: 339–43.

[16] Boudoulas H, Karayannacos PE, Lewis RP, Leier CV, Vasko JS. Effect of afterload on left ventricular performance in experimental animals. J Med 1982; 13: 373–85.

[17] De Cree J, Geukens H, Verhaegen H. A survey of 15 years experience with systolic time intervals. Acta Antwerpiensia 1987; 4: 2–18.

[18] De Cree, J Geukens H, Franken P, Verhaegen H. Non-invasive cardiac haemodynamics of nebivolol in men. Acta Antwerpiensia 1989; 6: 2–21.

[19] Garrard CL, Weissler AM, Dodge HT. The relationship of alterations in systolic time intervals to ejection fraction in patients wtih cardiac disease. Circulation 1970; 455–62.

[20] Stefadouros MA, Witham AC. Systolic time intervals by echocardiography. Circulation 1975; 51: 114–7.

[21] Adesanya CO, Sanderson JE, Verheijen ir PJT, Brinkman AW. Echocardiographic assessment and systolic time interval measurements in the evaluation of severe hypertension in Nigerian Africans. Aust NZ J Med 1981; 11: 364–9.

[22] Spodick DH, Doi YL, Bishop RL, Hashimoto T. Systolic time intervals reconsidered. Reevaluation of the preejection period: Absence of relation to heart rate. Am J Cardiol 1984; 53: 1667–70.

[23] Cokkinos DV, Heimonas ET, Demopoulos JN, Haralambakis A, Tsartsalis G, Gardikas CD. Influence of heart rate increase on uncorrected pre-ejection period/left ventricular ejection time (PEP/LVET) ratio in normal individuals. Br Heart J 1976; 38: 683–8.

[24] Boudoulas H, Lewis RP, Sherman JA, Bush CA, Daimangas G, Forester WF. Systolic time intervals in atrial fibrillation. Chest 1978; 74: 629–34.

[25] Boudoulas H, Geleris P, Bush CA *et al.* Assessment of ventricular function by combined noninvasive measures. Factors accounting for methodologic disparities. Int J Cardiol 1983; 2: 493–501.

[26] Weissler AM, Peeler RG, Roehil WH. Relationships between left ventricular ejection time, stroke volume, and heart rate in normal individuals and patients with cardiovascular disease. Am Heart J 1961; 62: 367–78.

[27] Greenfield JC, Harley A, Thompson HK, Wallace AG. Pressure-flow studies in man during atrial fibrillation. J Clin Invest 1968; 47: 2411–21.

[28] Harley A, Starmer CF, Greenfield JC. Pressure-flow studies in man. An evaluation of the duration of the phases of systole. J Clin Invest 1969; 48: 895–905.

[29] Tavel ME, Baugh DO, Feigenbaum H, Nasser W. Left ventricular ejection time in atrial fibrillation. Circulation 1972; 46: 744–52.

[30] Boudoulas H, Sohn YH, O'Neill W, Brown R, Weissler AM. The QT > QS_2 syndrome: a new mortality risk indicator in coronary artery disease. Am J Cardiol 1982; 50: 1229–35.

[31] Gillian RE, Parnes WP, Khan MA, Bouchard RJ, Warbasse JR. The prognostic value of systolic time intervals in angina pectoris patients. Circulation 1979; 60: 268–74.

[32] Northover BJ. Estimation of the risk of death during the first year after acute myocardial infarction from systolic time intervals during the first week. Br Heart J 1989; 62: 429–37.

[33] Weissler AM, O'Neill WW, Sohn YH, Stack RS, Chew PC, Reed AH. Prognostic significance of systolic time intervals after recovery from myocardial infarction. Am J Cardiol 1981; 48: 995–1002.

[34] Boudoulas H, Sohn YH, Brown R, Weissler AM. Left ventricular dysfunction is a determinant of improved survival with β-blockade therapy following myocardial infarction. IRCS J Med Sci 1985; 13: 271–2.

[35] Boudoulas H, Sohn YH, O'Neill WW, Brown R, Weissler AM. Identification of patients with improved survival following coronary bypass surgery. Cardiology 1984; 71: 247–54.

[36] Kyie MC, Freis ED. Serial measurements of systolic time intervals: Effects of propranolol alone and combined with other agents in hypertensive patients. Hypertension 1980; 2: 111–7.

[37] Boudoulas H, Mantzouratos D, Sohn YH, Weissler AM. Left ventricular mass and systolic performance in chronic systemic hypertension. Am J Cardiol 1986; 57: 232–7.

[38] Dodak A, Burg JR, Kloster FE. Systolic time intervals in chronic hypertension: alterations and response to treatment. Chest 1975; 68: 51–5.

[39] Geieris P, Boudoulas H, Schaal SF, Lewis R, Lima JJ. Effect of procainamide on left ventricular performance in patients with primary myocardial disease. Eur J Clin Pharmacol 1980; 18: 311–4.

[40] Boudoulas H, Beaver BM, Kates RE, Lewis RP. Pharmacodynamics of inotropic and chronotropic response to oral propranolol in normals and patients with angina. Chest 1978; 73: 146–9.

[41] Beiz GG, Erbel R, Schumann K, Gilfrich HJ. Dose-response relationships and plasma concentrations of digitalis glycosides in man. Eur J Clin Pharmacol 1978; 13: 103–11.

[42] Weissler AM, Schoenfeld CD. Effect of digitalis on systolic time intervals in heart failure. Am J Med Sci 1970; 259: 4–20.

[43] Lewis RP, Boudoulas H, Forester WF, Weissler AM. Shortening of electromechanical systole as a manifestation of excessive adrenergic stimulation in acute myocardial infarction. Circulation 1972; 46: 856–62.

[44] Toutouzas P, Gupta D, Sampon R, Shillingford J. Q-second sound interval in acute myocardial infarction. Br Heart J 1970; 32: 839–42.

[45] Boudoulas H, Geleris P, Lewis RP, Leier CV. Effect of increased adrenergic activity on the relationship between electrical and mechanical systole. Circulation 1981; 64: 28–33.

[46] Boudoulas H, Lewis RP, Vasko JS, Karayannacos PE, Beaver BM. Left ventricular function and adrenergic hyperactivity before and after saphenous vein bypass. Circulation 1976; 53: 802–6.

[47] Krukmeyer JJ, Boudoulas H, Binkley PF, Lima JJ. Comparison of hypersensitivity to adrenergic stimulation following abrupt withdrawal of propranolol and nadolol: influence of half-life differences. Am Heart J 1990; 120: 572–9.

[48] Boudoulas H, Lewis RP, Kates RE, Dalmangas G. Hypersensitivity to adrenergic stimulation after propranolol withdrawal in normal subjects. Ann Int Med 1977; 87: 433–6.
[49] Boudoulas H, Schaal SF, Lewis RP, Welch TG, DeGreen P, Kates RE. Negative inotropic effect of lidocaine in patients with coronary arterial disease and normal subjects. Chest 1977; 71: 170–5.
[50] Lima JJ, Boudoulas H, Blanford M. Concentration-dependence of disopyramide binding to plasma protein and its influence on kinetics and dynamics. J Pharm Exp Ther 1981; 219: 741–7.
[51] Lima JJ, Boudoulas H. Stereoselective effects of disopyramide in humans. J Cardiovasc Pharm 1987; 9: 594–9.
[52] Leier CV, Magorien RD, Boudoulas H, Lewis RP, Bambach D, Unverferth DV. The effect of vasodilator therapy on systolic and diastolic time intervals in congestive heart failure. Chest 1982; 81: 723–29.
[53] Cleland JGF, Dargie HJ, Ball SG *et al.* Effects of enalapril in heart failure: a double-blind study of exercise performance, renal function, hormones and metabolic state. Br Heart J 1985; 54: 305–12.
[54] Geleris P, Raidis C, Papadimitriou M, Boudoulas H, Metaxas P. Effect of hemodialysis on left ventricular performance. Am J Med 1983; 14: 211–22.
[55] Boudoulas H, Barrington W, Olson SM, Bashore TM, Wooley CF. Effect of acute standing and prolonged upright activity on left ventricular hemodynamics, systolic and diastolic intervals, and QT-QS_2 relationships. Am Heart J 1985; 110: 623–30.
[56] Van Der Hoeven GMA, Clerens PJA, Donders JJH, Beneken JEW, Vonk JTC. A study of systolic time intervals during uninterrupted exercise. Br Heart J 1977; 39: 242–53.
[57] Stafford RW, Harris WS, Weissler AM. Left ventricular systolic time intervals as indices of postural circulatory stress in man. Circulation 1970; 41: 485–92.
[58] Leighton RF, Weissler AM, Weinstein PB, Wooley CF. Right and left ventricular systolic time intervals. Am J Cardiol 1971; 27: 66–72.
[59] Grines CL, Bashore TM, Boudoulas H, Olson S, Shafer P, Wooley CF. Functional abnormalities in isolated left bundle branch block: The effect of interventricular asynchrony. Circulation 1989; 79: 845–53.
[60] Boudoulas H, Weinstein PB, Shaver JA, Wooley CF. Atrial septal defect: attenuation of respiratory variation in systolic and diastolic time intervals. J Am Coll Cardiol 1987; 9: 53–9.
[61] Boudoulas H, Mantzouratos D, Geleris P, Sohn Y, Lewis RP, Weissler AM. Hypertrophic cardiomyopathy: Noninvasive identification of patients with significant pressure gradient. Am J Noninvas Cardiol 1987; 1: 14–9.
[62] Boudoulas H, Geleris P, Lewis RP, Rittgers SE. The linear relationship between electrical systole. Mechanical systole, and heart rate. Chest 1981; 80: 613–7.

Clinical significance of systolic time intervals in hypertensive patients

M. HAMADA, K. HIWADA AND T. KOKUBU*

The 2nd Department of Internal Medicine, Ehime University School of Medicine, Ehime, Japan and
**The Department of Internal Medicine, Kinki Central Hospital (Itami), Hyogo, Japan*

KEY WORDS: Systolic time intervals, left ventricular hypertrophy, essential hypertension, pheochromocytoma, angina pectoris.

This paper updates the current view on clinical significance of systolic time intervals (STI) in estimating the cardiac changes associated with hypertension. The following three intervals were measured as STI: (1) electromechanical systole (QS_2 interval); (2) left ventricular ejection time (LVET) and (3) pre-ejection period (PEP). Firstly, the influences of changes in heart rate, preload, afterload and myocardial contractility upon each interval were reviewed; secondly, clinical applications of STI in various types of hypertension such as essential hypertension, hypertension with angina pectoris and pheochromocytoma were studied. In patients with essential hypertension, there was a good positive correlation between PEP and left ventricular mass, and a shortening of LVET was observed only at the decompensated stage. The changes in STI in angina pectoris with or without hypertension were similar and were different from those in essential hypertensives. STI in patients with pheochromocytoma were characterized by a marked shortening of QS_2 and LVET with normal PEP. These findings indicate the usefulness of STI in detecting cardiac changes in various types of hypertension.

Introduction

Systolic time intervals (STI) are determined non-invasively and reflect left ventricular function in a variety of disease states[1–14]. In addition, STI respond sensitively to exercise[15–18], postural change[19] and pharmaceutical agents[20–22]. However, use of STI in patients whose diagnosis is unclear leads to misinterpretation of STI[23]. For example, STI in hypertensives with and without coronary artery disease are markedly different[11]. Similarly, patients with angina pectoris with and without a history of myocardial infarction have different STI[10].

In this paper, we update the current view on the physiological meaning of each STI, and define changes in STI in various stages and types of hypertension.

Measurement of STI

As shown in Fig. 1, STI were measured from the simultaneous recording of an electrocardiogram (ECG), a phonocardiogram (PCG) and a carotid pulse trace (CPT) at a paper speed of 100 mm s^{-1}. Usually, the following STI were measured: (1) electromechanical systole QS_2, i.e. time from the beginning of the ECG QRS complex to the beginning of the aortic component of the second heart sound; (2) left ventricular ejection time (LVET), i.e. time from the onset of the upstroke on carotid pulse tracing to dicrotic notch. The initial upstroke and the end point of LVET were determined tangentially[10] as shown in Fig. 1; (3) pre-ejection period (PEP), i.e. $(QS_2) - ET$. These intervals were expressed as the mean of measurements over five consecutive beats. It is very important to identify accurately the initial point of the Q wave (Fig. 1) and to record clearly the high-frequency components of the second heart sound because the measurement of QS_2 is the basis of STI calculation.

Address for reprints: Mareomi Hamada, M.D., The 2nd Department of Internal Medicine, Ehime, University School of Medicine, Shigenobu, Onsen-gen, Ehime 791-02, Japan.

Pathophysiological significance of STI

QS_2 INTERVAL

The abbreviated QS_2 is related to increased adrenergic activity[24,25]. In fact, Lewis *et al.*[24] have reported a high correlation between the extent of shortening of QS_2 and the level of urinary catecholamine excretion in patients with acute myocardial infarction. Very recently we reported a

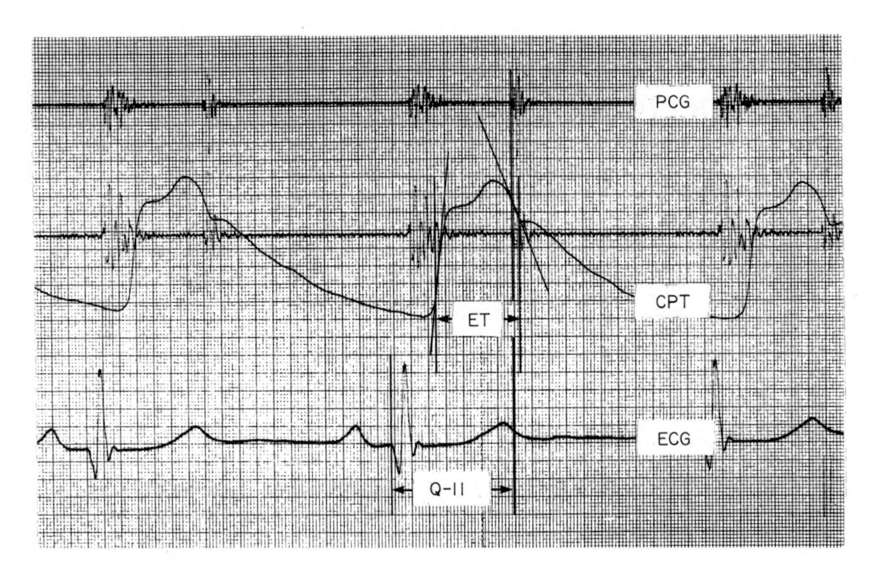

Figure 1 Diagram of the methods used for measuring systolic time intervals. PCG = phonocardiogram, CPT = carotid pulse tracing, ECG = electrocardiogram, ET = left ventricular ejection time (LVET).

marked shortening of QS_2 interval in patients with pheochromocytoma, even when asymptomatic[26]. This indicates that the measurement of STI can differentiate patients with pheochromocytoma from those with essential hypertension. The QS_2 interval may be a full index of adrenergic tone.

LVET

LVET usually correlates with stroke volume in an individual, but is greatly affected by heart rate, preload and myocardial contractility. Therefore, estimation of LVET must be careful, especially when LVET is compared between different individuals or groups. Fig. 2 shows the relationship between LVET and R-R interval in three normal controls and two patients with past myocardial infarction. R-R interval was changed by atrial pacing and there was an excellent positive correlation between LVET and R-R interval in all subjects. Left ventricular end-diastolic volume was not changed significantly by pacing vs rest, but left ventricular end-systolic volume was increased. Thus, stroke volume decreased on atrial pacing and LVET, which reflects stroke volume, shortened. Therefore, to compare LVET between different subjects with different R-R intervals, it is necessary to adjust LVET for heart rate.

Fig. 3 shows the changes in systolic blood pressure, heart rate, LVET index (corrected by the equation of Weissler *et al.*[2]), PEP index and left ventricular diastolic dimension (estimated by echocardiography) induced by skin application of 80 mg isosorbide dinitrate in patients with three normal controls and two patients with angina pectoris. Isosorbide dinitrate caused a marked decrease in preload, which was reflected by a decrease of left ventricular diastolic dimension that results in a reduction of stroke volume. This is responsible for a significant shortening of LVET or LVET index and a significant prolongation of PEP. We reported that LVET in patients with primary pulmonary hypertension and with atrial septal defect correlated very closely with left ventricular end-diastolic dimension[13,14].

Fig. 4 explains the relationship between LVET and myocardial contractility. Stroke index and LVET were determined at the same time each day, under no medication for at least 1 week. Cardiac output was measured by dye-dilution method using a cuvette densitometer. Heart rate showed no significant differences among the four patient groups studied. Patients with angina pectoris showed significantly longer LVET than controls. The latter had similar values of LVET to patients with past myocardial infarction without heart failure, but significantly greater mean systolic ejection rate, reflecting a difference in myocardial contractility. Thus, groups with similar LVET do not necessarily have the same stroke volume. Usually, a significant

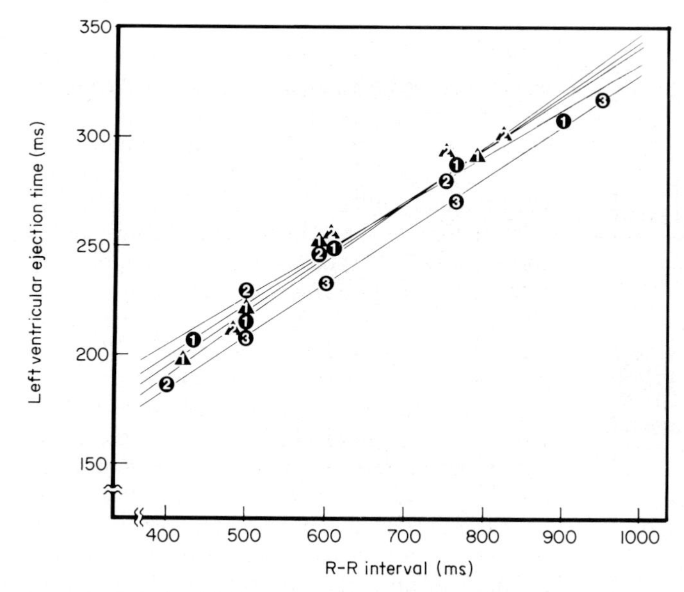

Figure 2 Relationship between R-R interval and left ventricular ejection time. ● = normal control, ▲ = past myocardial infarction.

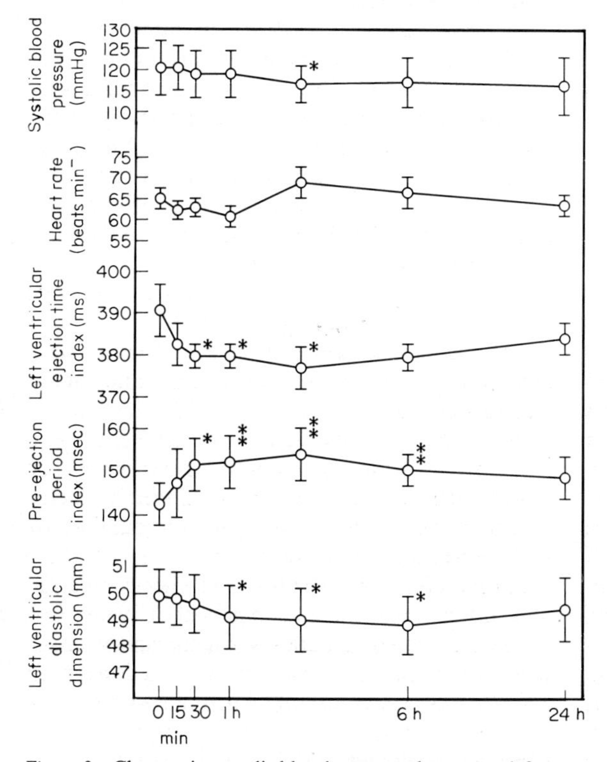

Figure 3 Changes in systolic blood pressure, heart rate, left ventricular ejection time index, pre-ejection period index and left ventricular diastolic dimension induced by skin application of isosorbide dinitrate. Data are mean ± SE. *$P < 0{\cdot}05$; **$P < 0{\cdot}02$.

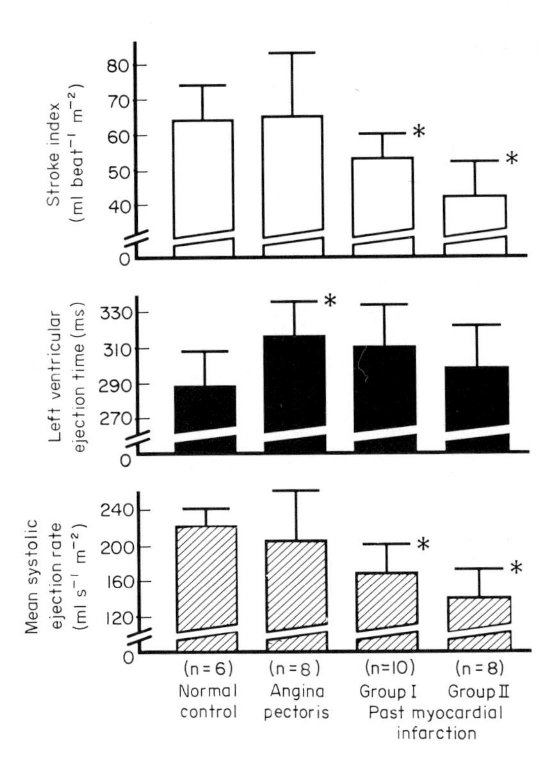

Figure 4 Comparisons of stroke index, left ventricular ejection time and mean systolic ejection rate among normal controls, patients with angina pectoris and patients with past myocardial infarction. Group I ejection fraction ⩾50%; Group II ejection fraction <50%.

shortening of LVET occurs only at the stage of overt heart failure[10,22].

Braunwald *et al.*[27] investigated the influence of afterload on the duration of left ventricular ejection using an isolated dog heart, and found that changes in mean aortic pressure had little influence on the duration of ejection. In the clinical setting, we compared LVET before and during isometric handgrip exercise in normal controls, patients with angina pectoris and patients with myocardial infarction without a history of heart failure. Despite a significant difference in ejection fraction, there were no significant differences in LVET or LVET index in any group[18]. Moreover, despite the significant elevation of blood pressure caused by isometric handgrip exercise, LVET showed no significant change and LVET index increased somewhat in all groups on exercise. Though these findings are due partly to the increase in catecholamines associated with exercise, a shortening of LVET does not ensue.

PEP

PEP is the summation of electrical and mechanical events between the onset of electrical depolarization of the left ventricle and the opening of the aortic valve. The change in electrical depolarization in various disease states is minimal, so the change in PEP mainly reflects a change in isovolumic contraction time, which is a good index of myocardial contractility. PEP is thought to depend upon three factors, namely, heart rate, preload and myocardial contractility. An early study on STI reported a significant negative relationship between PEP and heart rate[2,28], and the majority of studies using PEP in practice continued to correct PEP for heart rate. However, recent studies have shown that there is no significant correlation between PEP and heart rate[20,29–31]. Heart rate correction of PEP seems to be inappropriate in routine estimation of left ventricular function, because heart rate may itself have an effect, independent of the value of STI. However, when PEP is compared at marked different heart rates, e.g. during drug treatment or on exercise, it still remains to be determined whether heart rate correction is necessary.

As shown in Fig. 3, PEP was also markedly influenced by change in preload. There is a negative correlation between PEP and preload and, raising the legs, which increases preload, prolongs LVET and shortens PEP in controls and essential hypertensives[19]. Thus, an increase in stroke volume due to enhanced left ventricular filling is accompanied by lengthening of LVET and shortening of PEP and conversely, a decrease in stroke volume due to diminished diastolic filling is associated with abbreviated LVET and prolonged PEP.

The relationship between PEP and myocardial contractility has been investigated, and many studies have shown that a short PEP indicates a state of enhanced myocardial contractility. PEP is shortened by administration of epinephrine[21] or thyrotoxicosis,[4] and PEP is prolonged by beta-adrenergic blockade[20]. Therefore, the duration of PEP indicates the inotropic state.

Finally, great care must be taken when comparing data between inpatients and outpatients. Usually, PEP measured in hospital patients is significantly longer than in outpatients. This may be due to a decrease in adrenergic state associated with admission.

Clinical applications of STI

STI IN PATIENTS WITH ESSENTIAL HYPERTENSION

Fig. 5 shows the relationships between left ventricular hypertrophy and LVET or PEP in patients

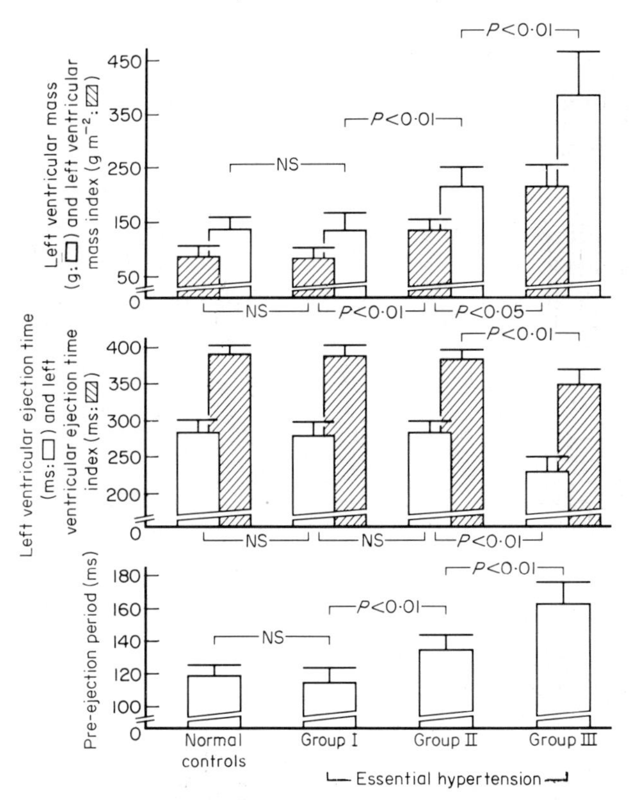

Figure 5 Comparisons of left ventricular mass, left ventricular mass index, left ventricular ejection time, left ventricular ejection time index and pre-ejection period among normal controls and three groups with essential hypertension.

with essential hypertension. Patients were subdivided into three groups based on interventricular septal thickness (IVST), left ventricular posterior wall thickness (PWT), and clinical symptoms: IVST and PWT were <10 mm in group I, IVST and/or PWT $\geqslant 10$ mm in group II and hypertensive patients in group III had a history of hypertensive heart failure. Left ventricular mass (LVM) was calculated by the method of Devereux and Reichek[32]. LVM index (LVMI) was derived by division of LVM by body surface area. LVMI was increased in the development of the stage of hypertension, however, neither ET nor LVET index showed significant differences between controls and group I or II essential hypertensive patients. A significant shortening of LVET or ET index was observed only in group III hypertensive patients with a history of heart failure. On the other hand, PEP was increased in proportion to the increase of LVM. As shown in Fig. 5, there was a good positive correlation between PEP and LVM in patients with essential hypertension. These results indicate first, that left ventricular hypertrophy associated with hypertension is almost always accompanied by a decrease in myocardial contractility, reflected in PEP prolongation and, second, that PEP itself is only an index of myocardial contractility in STI at a compensated stage of hypertension, because LVET was similar in controls, group I and II hypertensive patients. Only at a decompensated stage of hypertension, was the sensitivity of PEP slightly less than that of PEP/LVET ratio, which is considered a sensitive index of left ventricular contractility[3,22,33]. Very recently, we reported that STI in the decompensated stage of hypertension was not normalized by long-term antihypertensive therapy[34].

STI IN HYPERTENSIVE PATIENTS WITH ANGINA PECTORIS

It is generally considered that elevated blood pressure plays an important role in the pathogenesis of atherosclerosis. In fact, hypertension is one of

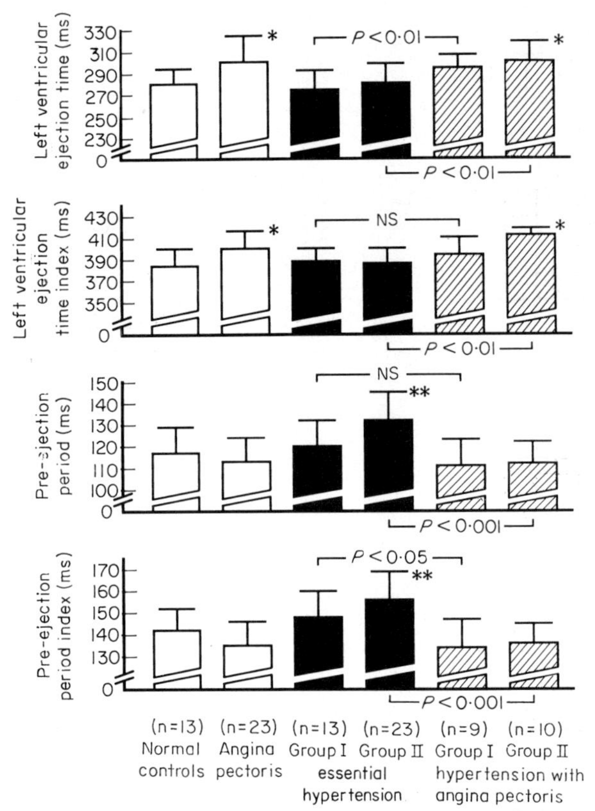

Figure 6 Comparisons of left ventricular ejection time, ejection time index, pre-ejection period and pre-ejection period index among normal controls, patients with angina pectoris, patients with essential hypertension and hypertensive patients with angina pectoris.

three major risk factors in coronary artery disease. In the clinical setting, we often encounter hypertensive patients with angina pectoris, and there is a significant difference in prognosis between those with and those without angina pectoris. To reduce the mortality rate in the former, it is very important to identify them effectively. Fig. 6 shows a comparison of STI between patients with angina pectoris, patients with essential hypertension and hypertensives with angina pectoris. All patients with angina pectoris had at least one coronary lesion of >75% stenosis confirmed by coronary cineangiography, but had no history of myocardial infarction. The group with hypertension was subdivided into two groups: group I without and group II with left ventricular hypertrophy. No drugs were administered during the study. We have reported that STI in patients with angina pectoris without a high-risk coronary artery lesion were characterized by LVET prolongation with normal PEP[10]. In patients with a high-risk lesion AHA class 5/6[35], STI were characterized by shortening of LVET and normal PEP. As shown in Fig. 6, in hypertensive patients with angina pectoris, LVET or LVET index were prolonged and PEP was normal, indicating that the changes in STI in essential hypertension with angina pectoris are different from those in patients with essential hypertension, but similar to those in patients with angina pectoris[11]. The prolongation of LVET in hypertensives with angina pectoris might be the consequence of reduced myocardial contractility due to reduced blood supply through the stenotic coronary artery, in other words, a decrease of systolic ejection rate, while the relative shortening of PEP in hypertensive patients with angina pectoris might be caused by a high level of circulating catecholamines.

STI IN PATIENTS WITH PHEOCHROMOCYTOMA

Early diagnosis of pheochromocytoma is very important because it is potentially fatal if un-

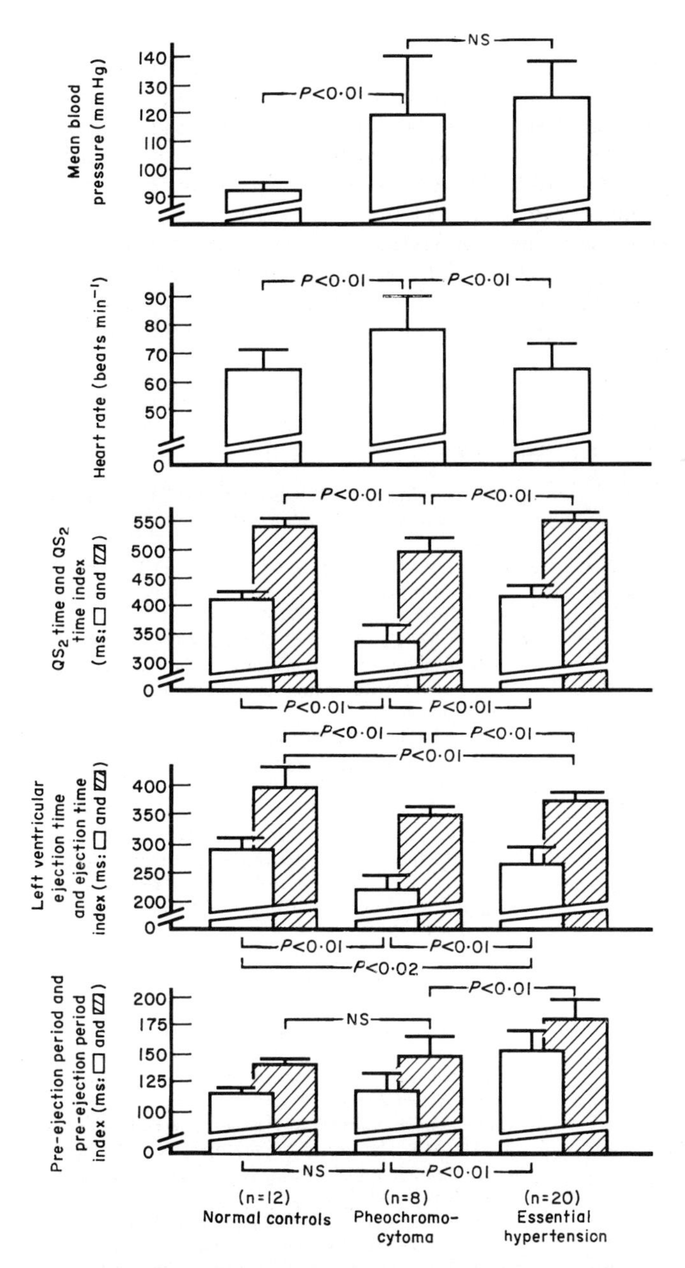

Figure 7 Comparisons of mean blood pressure, heart rate, QS_2 and QS_2 index, left ventricular ejection time and ejection time index, pre-ejection period and pre-ejection period index among normal controls, patients with pheochromocytoma and patients with essential hypertension.

detected. Very recently, we reported that STI could be a good screening procedure for detecting pheochromocytoma[26]. Fig. 7 shows STI from eight patients with pheochromocytoma compared with those from controls and patients with essential hypertension. In all groups data were obtained at least 1 week after medication was discontinued. Marked shortening of QS_2 or QS_2 index and LVET or LVET index were common characteristics in patients with pheochromocytoma. On the other hand, PEP or PEP index showed no significant difference between patients with pheochromocytoma and normal controls. The shortening of QS_2 or QS_2 index in pheochromocytoma was so marked that no overlap in the value of QS_2 or QS_2 index was observed among three groups. The markedly abbreviated QS_2 in patients with pheochromocytoma is due to the rise of adrenergic activity[24].

Since many patients with decompensation associated with hypertension were included in the essential hypertension group, LVET or LVET index was significantly shorter in this group than in controls. However, LVET or LVET index was much shorter in patients with pheochromocytoma than in patients with essential hypertension, and comparable to that in the decompensated hypertensive subgroup. Shortening of LVET in the former was not necessarily due to a reduced R-R interval, but probably to chronically raised catecholamines in pheochromocytoma. Fig. 1 is a recording from a patient with pheochromocytoma: heart rate was in the normal range (67 beats min^{-1}), but LVET was markedly shortened (240 ms). Though the exact mechanism for the marked shortening of LVET remains to be determined, the following mechanisms are suggested: (1) increased peripheral vasoconstriction associated with α-adrenergic stimulation[36]; (2) myocardial damage by excess β-adrenergic stimulation, so-called catecholamine myocarditis[37,38]. Recent studies suggest that the cardiomyopathy is mediated mainly by the α-adrenergic receptor system[39]. One patient with pheochromocytoma had a significant elevation of cardiac enzymes associated with excess catecholamines during a hypertensive period; (3) significant reduction in the total blood volume in patients with pheochromocytoma[40–42]. In this condition, decreased stroke volume and shortened ET related to a decrease in venous return may be produced.

Normal PEP in patients with pheochromocytoma is also markedly different from that in hypertensives. The changes in LVET and PEP observed in patients with pheochromocytoma are different from those following administration of epinephrine or norepinephrine in normal controls. On administration of epinephrine, PEP shortened dose-dependently and LVET index remained unchanged[20,21]. These findings indicate that myocardial adaptation to short-term administration of catecholamines and to longterm catecholamine excess due to pheochromocytoma are markedly different.

In conclusion, different from the indices in diastolic phase, STI cannot detect the abnormalities in cardiac function in hypertensive patients without left ventricular hypertrophy, but STI can detect very sensitively the disturbance of cardiac function in moderate to advanced stages of hypertension. In addition, STI are the only method to estimate the adrenergic state at any given moment in the clinical setting. STI are thought to be one of the most sensitive and reliable methods to evaluate cardiac status at various stages and in various types of hypertension.

References

[1] Weissler AM, Harris WS, Schoenfeld CD. Systolic time intervals in heart failure in man. Circulation 1968; 37: 149–59.

[2] Weissler AM, Harris WS, Schoenfeld CD. Bedside technics for the evaluation of ventricular function in man. Am J Cardiol 1969; 23: 577–83.

[3] Ahmed SS, Levinson GE, Schwartz CJ, Ettinger PO. Systolic time intervals as measures of the contractile state of the left ventricular myocardium in man. Circulation 1972; 46: 559–71.

[4] Parisi AF, Hamilton BP, Thomas CN, Mazzaferri EL. The short cardiac pre-ejection period. An index to thyrotoxicosis. Circulation 1974; 49: 900–4.

[5] Meng R, Hollander C, Liebson PR, Teran JC, Barresi V, Lurie M. The use of noninvasive methods in the evaluation of left ventricular performance in coronary artery disease. I. Relation of systolic time intervals to angiographic assessment of coronary artery disease severity. Am Heart J 1975; 90: 134–44.

[6] Lewis RP, Boudoulas H, Welch TG, Forester WF. Usefulness of systolic time intervals in coronary artery disease. Am J Cardiol 1976; 37: 787–96.

[7] Stack RS, Lee CC, Reddy BP, Taylor ML, Weissler AM. Left ventricular performance in coronary artery disease evaluated with systolic time intervals and echocardiography. Am J Cardiol 1976; 37: 331–9.

[8] Eddleman EE, Swatzell RH, Bancroft WH, Baldone JC, Tucker MS. The use of the systolic time intervals for predicting left ventricular ejection fraction in ischemic heart disease. Am Heart J 1977; 93: 450–4.

[9] Gillian RE, Parnes WP, Khan MA, Bouchard RJ, Warbasse JR. The prognostic value of systolic time intervals in angina pectoris patients. Circulation 1979; 60: 268–75.

[10] Hamada M, Kazatani Y, Shigematsu Y *et al.* Clinical significance of systolic time intervals for the evaluation of left ventricular function in patients with coronary artery disease. Jpn Circ J 1983; 47: 810–6.

[11] Ochi T, Hamada M, Kazatanai Y *et al.* Noninvasive evaluation of left ventricular function by systolic time intervals in essential hypertension with angina pectoris. Jpn Circ J 1984; 48: 1299–305.

[12] Ito T, Hamada M, Shigematsu Y *et al.* The analysis of systolic and diastolic time intervals: A more sensitive non-invasive method in the assessment of left ventricular

dysfunction in the patients with essential hypertension. Clin Exp Hypertens 1985; A7: 951–63.
[13] Shigematsu Y, Hamada M, Ito T, Kokubu T. Non-invasive assessment of left ventricular function in patients with primary pulmonary hypertension and atrial septal defect. J Cardiovasc Ultrason 1986; 5: 63–9.
[14] Shigematsu Y, Hamada M, Kokubu T. Significance of systolic time intervals in predicting prognosis of primary pulmonary hypertension. J Cardiol 1988; 18: 1109–14.
[15] Pouget JM, Harris WS, Mayron BR, Naughton JP. Abnormal responses of the systolic time intervals to exercise in patients with angina pectoris. Circulation 1971; 43: 289–98.
[16] McConahay DR, Martin CM, Cheitlin MD. Resting and exercise systolic time intervals. Correlations with ventricular performance in patients with coronary artery disease. Circulation 1972; 45: 592–601.
[17] Gillian RE, Parues WP, Mondeel BE, Bouchard RJ, Warbasse JR. Systolic time intervals before and after maximal exercise treadmill testing for evaluation of chest pain. Chest 1977; 71: 479–85.
[18] Hamada M. Estimation of left ventricular functional reserve in patients with coronary artery disease by isometric handgrip exercise (with English abstract). J Juzen Med Soc 1978; 87: 774–89.
[19] Mukai M, Hamada M, Sumimoto T, Sekiya M, Kokubu T. Disparate difference in preload reserve between myocardial hypertrophy due to essential hypertension and hypertrophic cardiomyopathy. J Hypertens 1988; 6 (Suppl 4): S138–40.
[20] Harris WS, Schoenfeld CD, Weissler AM. Effect of adrenergic receptor activation and blockade on the systolic preejection period, heart rate and arterial pressure in man. J Clin Invest 1967; 46: 1704–14.
[21] Salzman SH, Wolfson S, Jackson B, Schechter E. Epinephrine infusion in man. Standardization, normal response and abnormal response in idiopathic hypertrophic subaortic stenosis. Circulation 1971; 43: 137–44.
[22] Cleland JGF, Dargie HJ, Ball SG *et al.* Effects of enalapril in heart failure: a double-blind study of exercise performance, renal function, hormones, and metabolic state. Br Heart J 1985; 54: 305–12.
[23] Parker ME, Just HG. Systolic time intervals in coronary artery disease as indices of left ventricular function: Fact or fancy? Br Heart J 1974; 36: 368–76.
[24] Lewis RP, Boudoulas H, Forester WF, Weissler AM. Shortening of electromechanical systole as a manifestation of excessive adrenergic stimulation in acute myocardial infarction. Circulation 1972; 46: 856–62.
[25] Boudoulas H, Lewis RP, Vasko JS, Karayannacos PE, Beaver BM. Left ventricular function and adrenergic hyperactivity before and after saphenous vein bypass. Circulation 1976; 53: 802–6.
[26] Hamada M, Ito T, Hiwada K, Kokubu T, Genda A, Takeda R. Hemodynamic characteristics in patients with pheochromocytoma: Systolic time intervals aid the detection of pheochromocytoma. Jpn Circ J (in press).
[27] Braunwald E, Sarnoff SJ, Stainsby WN. Determinants of duration and mean rate of ventricular ejection. Circ Res 1958; 6: 319–25.
[28] Weissler AM. Current concepts in cardiology. Systolic time intervals. N Engl J Med 1977; 296: 321–4.
[29] Cokkinos DV, Heimonas ET, Demopoulos JN, Haralambakis A, Tsartsalis G, Gardikas CD. Influence of heart rate increase on uncorrected pre-ejection period/left ventricular ejection time (PEP/LVET) ratio in normal individuals. Br Heart J 1976; 38: 683–8.
[30] Spodick DH, Doi YL, Bishop RL, Hashimoto T. Systolic time intervals reconsidered. Reevaluation of the preejection period: Absence of relation to heart rate. Am J Cardiol 1984; 53: 1667–70.
[31] Sundberg S. Influence of heart rate on systolic time intervals. Am J Cardiol 1986; 58: 1144–5.
[32] Devereux RB, Reichek N. Echocardiographic determination of left ventricular mass in man. Anatomic validation of the method. Circulation 1977; 55: 613–8.
[33] Garrard CL, Weissler AM, Dodge HT. The relationship of alterations in systolic time intervals to ejection fraction in patients with cardiac disease. Circulation 1970; 42: 455–62.
[34] Hamada M, Shigematsu Y, Suzuki M *et al.* Does left ventricular systolic dysfunction in hypertensive patients with heart failure return to normal by chronic antihypertensive therapy? Am J Cardiol (in press).
[35] Austen WG, Edwards JE, Frye RL *et al.* AHA committee report: A reporting system on patients evaluated for coronary artery disease. Circulation 1975; 51: 5–40 (News from the American Heart Association).
[36] Messerli FH, Finn M, MacPhee AA. Pheochromocytoma of the urinary bladder: Systemic hemodynamics and circulating catecholamine levels. J Am Med Assoc 1982; 247: 1863–4.
[37] Szakacs JE, Cannon A. L-Norepinephrine myocarditis. Am J Clin Pathol 1958; 30: 425–34.
[38] Kline IK. Myocardial alterations associated with pheochromocytoma. Am J Pathol 1961; 38: 539–51.
[39] Lee JC, Sponenberg DP. Role of β-adrenoceptors in norepinephrine-induced cardiomyopathy. Am J Pathol 1985; 121: 316–21.
[40] Brunjes S, Johns VJ Jr, Crane MG. Pheochromocytoma. Postoperative shock and blood volume. N Engl J Med 1960; 262: 393–6.
[41] Tarazi RC, Dustan HP, Frohlich ED, Gifford RW, Hoffman GC. Plasma volume and chronic hypertension. Relationship to arterial pressure levels in different hypertensive disease. Arch Intern Med 1970; 125: 835–42.
[42] Deoreo GA Jr, Stewart BH, Tarazi RC, Gifford RW. Preoperative blood transfusion in the safe surgical management of pheochromocytoma: A review of 46 cases. J Urol 1974; 111: 715–21.

Systolic time intervals as indicators for cardiac function in rat models for heart failure

R. G. SCHOEMAKER AND J. F. M. SMITS

Department of Pharmacology, University of Limburg, Maastricht, The Netherlands

KEY WORDS: Myocardial infarction, hypertension, stroke volume, peripheral resistance, pre-ejection period, left ventricular ejection time, inotropes, vasodilators.

Measurement of the pre-ejection period (PEP) can give valuable information on myocardial function in rats. Depressed cardiac function is accompanied by prolongation of PEP, which shortens when cardiac function improves. Left ventricular ejection time (LVET) may be shortened with ageing in spontaneously hypertensive rats.

Introduction

Systolic time intervals (STI) are gaining acceptance as indicators for cardiac function in clinical practice. They can be obtained non-invasively and without discomfort for the patients.

In animal studies, the use of systolic time intervals as indicators for cardiac function is rare. Nevertheless, their measurement is relatively easy and might provide useful information on the condition of the myocardium. We have devised methods for concomitant measurement of cardiac function and systolic time intervals in rats with varying, documented degrees of heart failure, due to either long-lasting spontaneous hypertension or myocardial infarction. Here, we compare systolic time intervals (STI) and cardiac function in order to verify the usefulness of the former in physiological studies in the rat. Furthermore, our results may provide a validation of the use of STI as indicators of cardiac function in other species, including man.

Animal models

Several models have been developed for the study of pathophysiology and therapy of heart failure[1]. Of these, the two explored most extensively are the rat post-myocardial infarction[2–4], and the ageing spontaneously hypertensive rat (SHR)[5,6]. Since it has been our belief, from the onset of our studies, that the best information can be derived from animals that have all haemodynamic control mechanisms fully operational, we developed techniques to allow haemodynamic assessment in conscious, unrestrained animals.

The haemodynamic variables that are routinely measured are mean arterial pressure (MAP), heart rate (HR), central venous pressure (CVP), cardiac

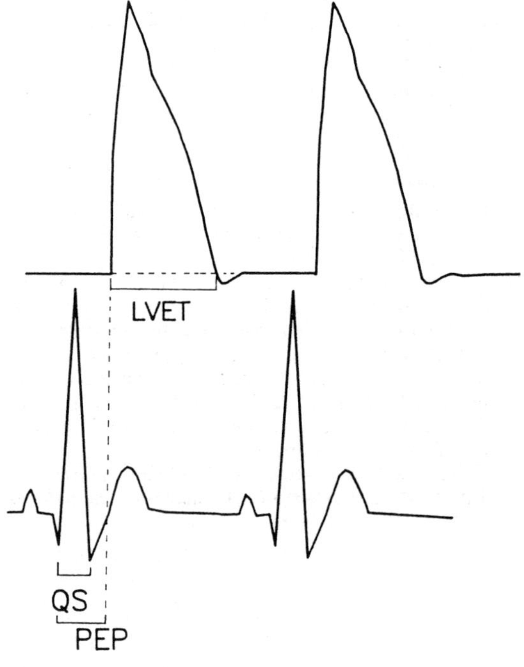

Figure 1 Scheme illustrating the temporal relationships between ECG and electromagnetic flow (EMF) signal obtained on the ascending aorta.

J. F. M. Smits is an Established Investigator for the Dutch Heart Foundation.

Correspondence: Jos F. M. Smits, Department of Pharmacology, University of Limburg, PO Box 616, Maastricht, The Netherlands.

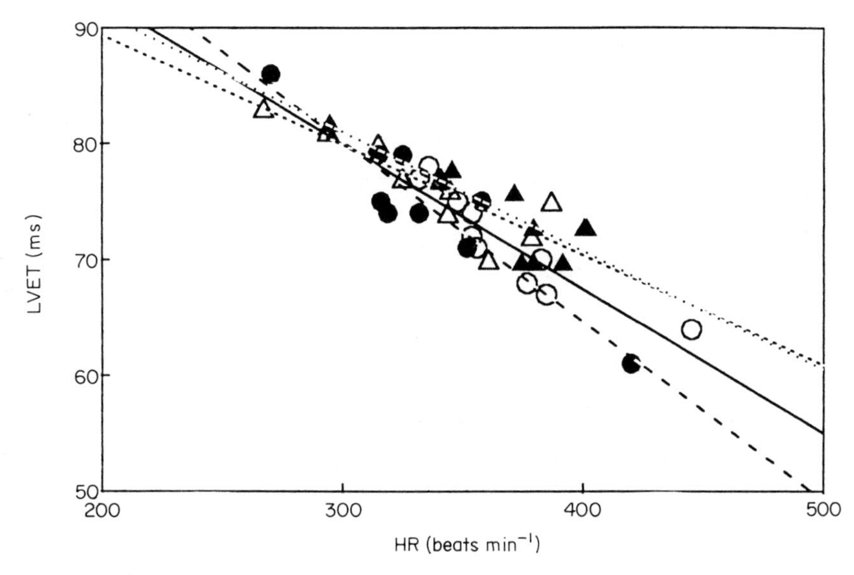

Figure 2 Relationship between LVET and HR in groups of normotensive and spontaneously hypertensive rats. (data from ref. 9). ○ = SHR 3 months; ● = SHR 12 months; △ = SHR 17 months; ▲ = WKY 17 months.

Table 1 Evolution of systolic time intervals during ageing in spontaneously hypertensive rats. All values are means ± SEM; all intervals are in ms

Strain: Age:	SHR 3 months	SHR 12 months	SHR 17 months	WKY 17 months	SHR 20 months
Parameter					
N	10	10	10	11	5
QS	17±1	18±1	18±1^{+}	14±1	17±2
PEP	35±1	36±1	42±1$^{+,*}$	35±1	42±1*
EMST	107±2	111±2	118±2$^{+,*}$	108±2	113±2
LVET	72±1	75±2	77±1$^{+,*}$	74±1	71±2
LVETI	118±2	127±2*	103±1$^{+,*}$	113±2	ND
PEP/LVET	0·50±0·02	0·50±0·03	0·55±0·02$^{+,*}$	0·48±0·07	0·59±0·02*
PEP/LVETI	0·30±0·03	0·29±0·03	0·42±0·04$^{+,*}$	0·30±0·03	ND

$^{+}P<0{\cdot}05$ vs age-matched WKY (tested only at 17 months); $^{*}P<0{\cdot}05$ vs 3-month-old SHR. ND = not determined.

output (CO) and the ECG. CO is measured directly with an electromagnetic flow probe that is implanted on the ascending aorta, approximately 2 mm above the heart[7,8]. Our methods allow continuous recording of all signals. From combinations of signals other variables can be calculated, such as stroke volume and total peripheral resistance. Systolic time intervals are measured from the ECG, the flow-signal or a combination of both; QS width and left ventricular ejection time (LVET) can be obtained directly from the ECG and the flow-signal, respectively. As illustrated in Fig. 1, pre-ejection period (PEP) can be measured as the interval between the Q-deflection in the ECG and the start of ejection. Simple addition of PEP and LVET yields the total time for electromechanical systole (EMST), the equivalent of QS_2 in man. For reliable sampling, we developed a computer system that allows continuous, beat-to-beat sampling and 'on-line' calculation of all derived variables at a rate of 500 Hz. In the studies discussed below, they were averaged over 1 min periods. Typically, within

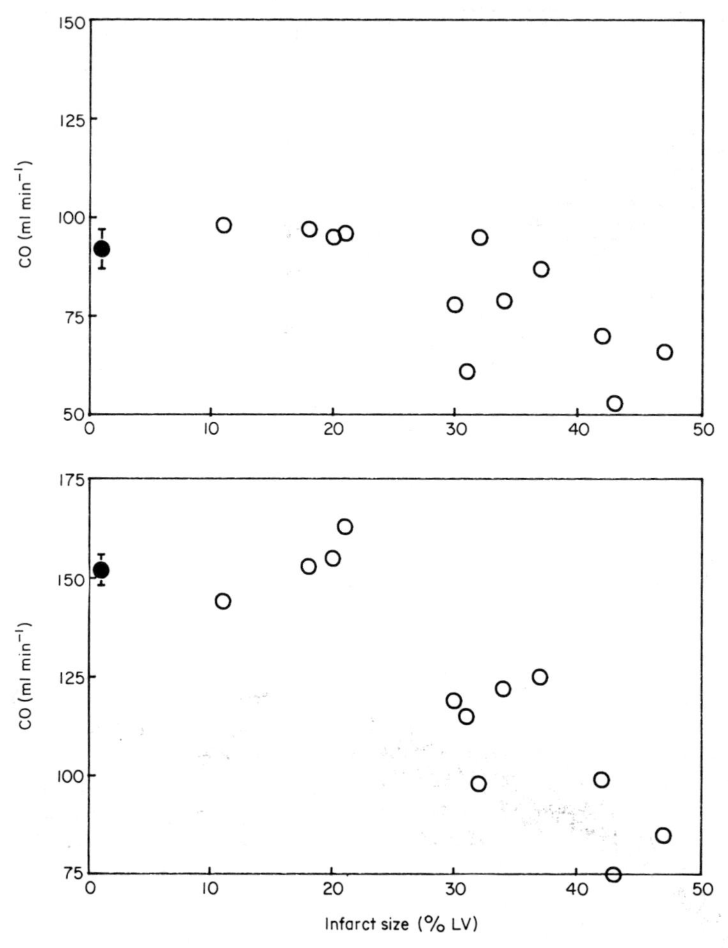

Figure 3 Cardiac output at rest (top) and upon volume loading (bottom) in rats with different infarct sizes, 5 weeks after infarction. The data-point at 0% represents the mean ± SEM for nine sham-operated animals.

groups, coefficients of variation for QS, PEP, LVET and EMST are 15%, 10%, 6% and 4·5%, respectively. The gradual decrease of the coefficient of variation with increasing interval time may depend upon the resolution of our measuring system, which, by virtue of its sampling rate, is 2 ms.

In heart failure, the cardiac function curve relating cardiac output to filling pressures is, by definition, depressed. In man this implies that maximal cardiac output obtained during exercise is depressed, while resting cardiac output may still be normal. In rats with heart failure following coronary artery ligation, exercise CO has been shown to be similarly depressed[2]. However, it was unpractical to subject our instrumented rats to exercise. Therefore, the method of Pfeffer and co-workers devised to obtain cardiac function curves in anaesthetized open chest rats[4] was adopted for use in our conscious animals. This method employs rapid volume loading with an isotonic Ringer's solution such that 12 ml of Ringer's solution (at 37 °C) is infused over a 1-min period, while continuously monitoring CVP and CO. This method yields extremely reproducible cardiac function curves[9,11]. Baseline and stimulated CO are then used as indicators of overall cardiac function.

STI in relation to cardiac function in ageing SHR

Previous studies in anaesthetized open-chest animals[5,6,12–15] have suggested that SHR develop heart failure similar to that observed in patients with long-standing hypertension. In these studies, heart failure was overt from 18 months of age. In

Table 2 Evolution of STI following coronary artery ligation in rats. Data are means ± SEM; all intervals are in ms

Parameter	Group	Time	
		3 weeks	5 weeks
N	sham	8	7
	infarcted	9	8
QS	sham	11±1	16±1
	infarcted	15±2	17±2
PEP	sham	36±1	36±1
	infarcted	41±2*	40±1*
EMST	sham	111±2	111±2
	infarcted	117±4	115±2
LVET	sham	74±1	75±1
	infarcted	76±1	75±2
PEP/LVET	sham	0·48±0·01	0·49±0·01
	infarcted	0·52±0·02	0·54±0·02*

*Significantly different from sham-operated animals.

our experiments, haemodynamic measurements were obtained in SHR at 3, 12, and 17 months of age, and in normotensive Wistar Kyoto (WKY) rats at 3 and 17 months of age. Because surgery was hampered by the bad condition of the animals, measurements at 20 months were obtained in only a small group of SHR.

MAP and total peripheral resistance (TPR) were consistently elevated in SHR compared with WKY. Up to 17 months, ageing did not affect these variables. Both resting and stimulated CO were similar in the two groups at all ages[9,16].

STI did show changes over time in SHR. In man, LVET and QS_2 have been shown to be related to HR[17–19]. Also, PEP has been suggested to be related to HR[17,18,20], but other studies could not confirm this[21,22]. In order to examine a possible relationship between HR and STI in our SHR and WKY, linear regression analyses were performed on these measurements. Rather than manipulating HR in separate animals, we chose to do regression analyses on the population. In man, this has been shown to be a good alternative to individual correlations[20]. Because of strain- and age-differences these regressions were carried out on the different subpopulations. Although there was a complete lack of correlation between HR and PEP, QS-width and EMST, LVET was found to be highly significantly correlated to HR in all subgroups (Fig. 2). Therefore, only LVET was indexed for a theoretical HR of 0 beats min^{-1} (LVETI).

Evolution of PEP, LVET, EMST, LVETI, PEP/LVET and PEP/LVETI with ageing in SHR is shown in Table 1. All measured intervals increased with age in SHR, up to an age of 17 months. At that age, they were all significantly elevated compared with both 3-month-old SHR and 17-month-old WKY. LVETI showed an increase at 12 months, but was suppressed at 17 months. At 20 months, LVET had decreased again in SHR, whereas PEP had not changed. The calculated ratios PEP/LVET and PEP/LVETI also increased significantly, and the increase in PEP/LVET continued up to 20 months.

PEP is dependent upon ventricular conductance time and the rate of isometric contraction[19]. Conductance is reflected in QS-width, and the prolonged QS-width in SHR vs WKY at 17 months may contribute to the prolonged PEP in the former strain. However, it does not explain the gradual increase in PEP within the SHR strain, which must depend upon a decreased rate of contraction. This is consistent with the shift of myosin isoenzymes from the fast V_1-type to the slower V_3-type[23–25]. Thus, from the prolongation of PEP we conclude that a

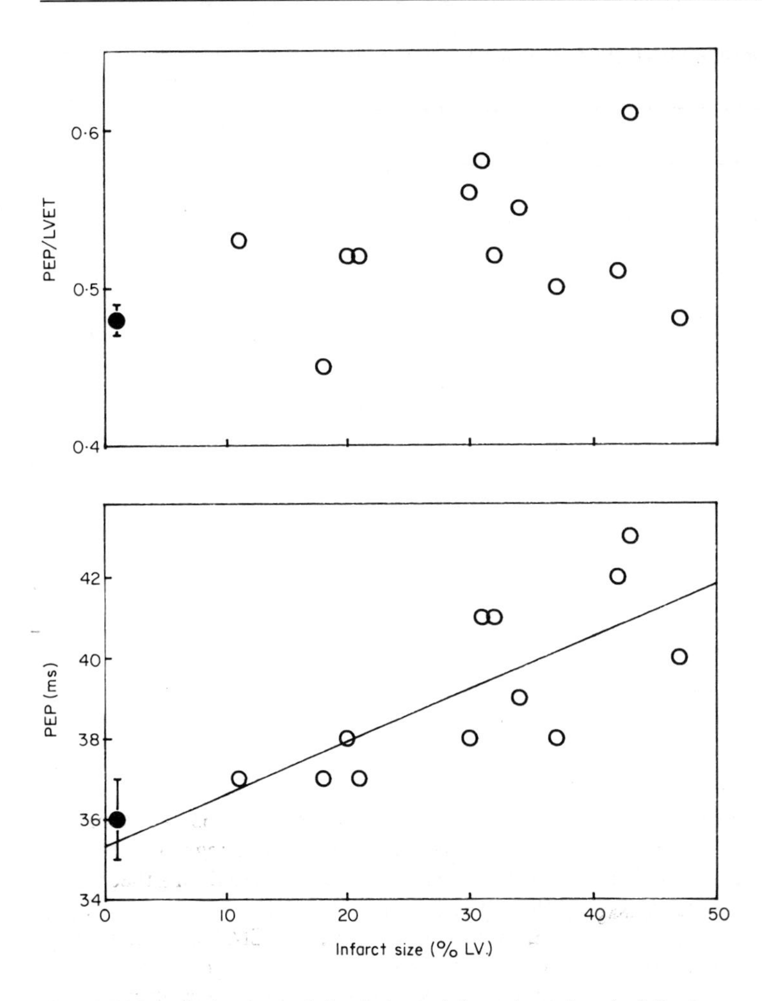

Figure 4 Systolic time intervals in relation to infarct size at 5 weeks following infarction. The correlation between infarct size and PEP was significant ($r=0{\cdot}8$; $P<0{\cdot}01$). The data-point at 0% represents the mean ± SEM for nine sham-operated animals.

gradual change occurs in the myocardium that is consistent with gradually decreasing contractility of the hypertrophied myocardium.

The ratio between PEP and LVET has been shown to correlate closely with ejection fraction in man[22,26]. The increase in PEP/LVET with age that we observed in SHR lends further support to the presence of heart failure in this model. In separate groups of anaesthetized SHR and WKY we measured ejection fraction with multigated equilibrium cardiac blood pool imaging with autologous 99 m technetium-labelled red blood cells. We observed that ejection fractions at 20 months were similar in SHR ($65 \pm 3\%$; N=5) and WKY ($61 \pm 4\%$; N=5). At 22 months, ejection fraction had slightly increased in WKY ($70 \pm 2\%$; N=6), whereas it had decreased to $47 \pm 4\%$ (N=6) in SHR. Although we did not measure STI in the latter groups of animals, the fact that at 20 months PEP/LVET was increased, despite similar ejection fractions in a group of animals of similar age, we conclude that there is no relationship between PEP/LVET or PEP/LVETI and ejection fraction in this model of heart failure.

STI in heart failure following myocardial infarction in rats

The second well-explored model for heart failure in rats is based on myocardial infarction following

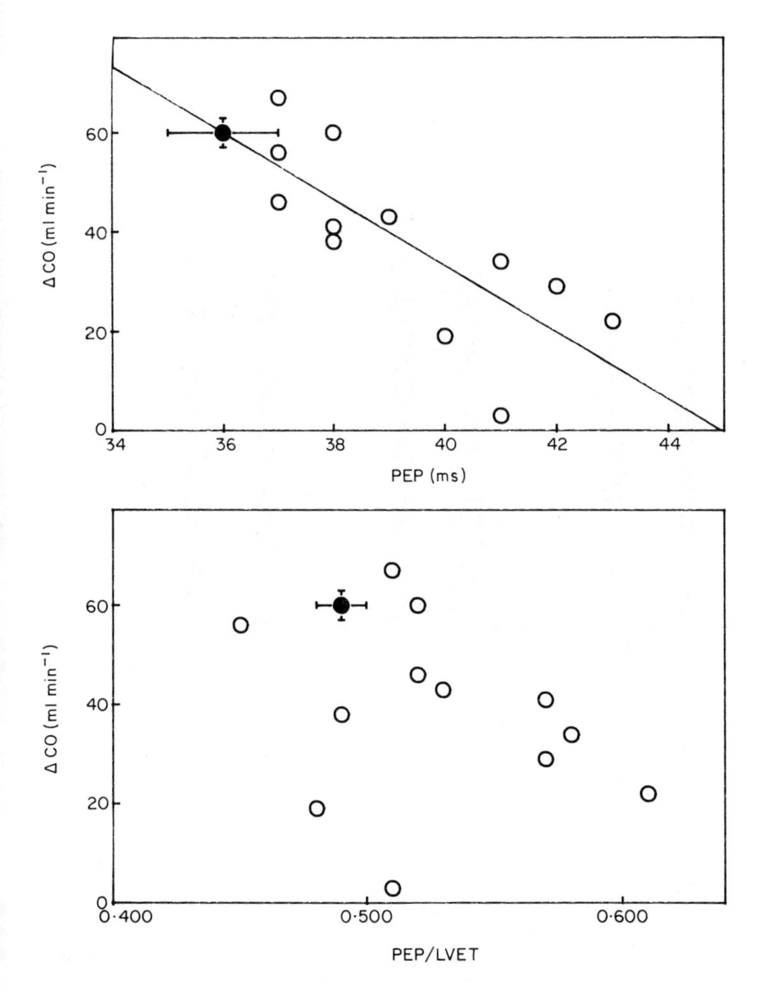

Figure 5 Relationship between systolic time intervals and the effect of volume loading on cardiac output; the correlation between PEP and ΔCO was significant (r = −0·8; $P<0{\cdot}01$). Closed symbols represent mean ± SEM for nine sham-operated animals.

ligation of the left descending coronary artery[2–4]. Generally, the period following infarction is subdivided into two phases. The first 3 weeks following infarction are assumed to be the period in which the infarcted area is replaced with scar tissue. Thereafter occurs a period during which failure gradually develops[27]. We have taken this distinction into account in our studies, most of which have been performed at 5 weeks after infarction, i.e. well after the healing period.

Ligation of the coronary artery in rats results in infarction of 10–50% of the total left ventricular wall in survivors. The effect of such infarcts on cardiac function at 5 weeks after the surgery is shown in Fig. 3. Both resting and stimulated CO were related to infarct size only for infarcts greater than 25% of the left ventricle, suggesting that smaller infarcts can be compensated for completely. When we excluded results from animals with infarcts ≤25%, stimulated and, to a lesser degree, resting CO significantly decreased following infarction. Resting MAP, CVP and HR were no different in infarcted and sham-operated rats[9,28,29].

Mean values for STI at 3 and 5 weeks following infarction of more than 25% or sham surgery are summarized in Table 2. PEP increased significantly at both times. LVET was slightly prolonged at 3 weeks, but not at 5 weeks. EMST did not change

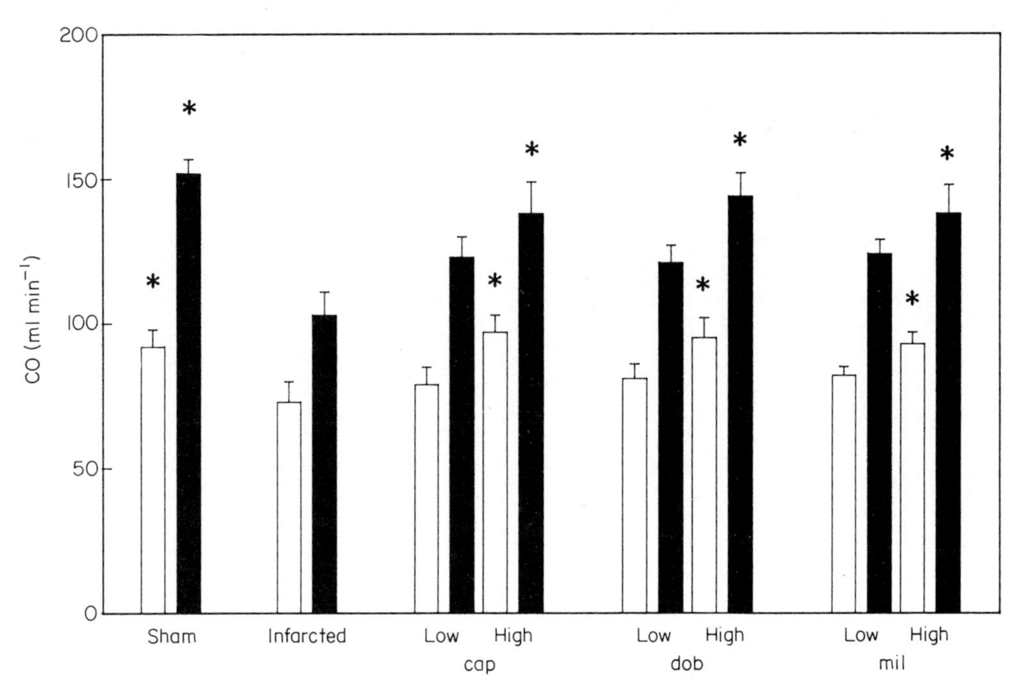

Figure 6 Resting (open bars) and stimulated CO (black bars) following 2-week treatment with captopril (cap), milrinone (mil) and dobutamine (dob). 'low' and 'high' refer to low and high doses for the drugs, mentioned in the text. * = significantly different ($P<0{\cdot}05$) from untreated infarcted animals.

significantly, and PEP/LVET showed a non-significant increase at 3 weeks, and significant elevation at 5 weeks.

The relationship between infarct size and PEP and PEP/LVET at 5 weeks is illustrated in Fig. 4. There was a highly significant linear relationship between PEP and infarct size. Although PEP/LVET tended to increase with infarct size, linear regression analysis did not yield a significant correlation. However, for infarcts greater than 25% of the left ventricle, PEP/LVET was significantly elevated in infarcted vs sham animals (PEP/LVET; $0{\cdot}49 \pm 0{\cdot}01$ vs $0{\cdot}54 \pm 0{\cdot}02$; $P<0{\cdot}01$). Since resting and stimulated CO in these animals were depressed, PEP/LVET might be an indicator for overall cardiac function in these animals.

To investigate this further, we compared resting PEP and PEP/LVET with CO during voume loading, as shown in Fig. 5. Again, the relationship between PEP and this indicator of cardiac function was linear, whereas there was no correlation between PEP/LVET and the response to volume loading.

Since QS-width was not affected by the infarction, increased PEP again indicates decreased contractility, corroborating previous work in this model[2,4,30]. That the relationship between measured infarct size and PEP was linear, and that PEP was prolonged in all animals with a myocardial infarct, suggest that PEP may be the most sensitive indicator for dysfunction in this model of heart failure.

Effects of chronic therapy following myocardial infarction

We studied the effects of three different drugs, milrinone, dobutamine and captopril, on cardiac function and STI following myocardial infarction in rats. All three have a documented beneficial haemodynamic effect in human heart failure. Both milrinone and captopril were given continuously with osmotic minipumps 3–5 weeks after infarction. Infusions were continued during the experiments, at doses of 30 or 150 g kg^{-1} h^{-1} for milrinone, and 0·1 or 0·5 mg kg^{-1} h^{-1} for captopril. Dobutamine was given in two daily intraperitoneal injections of $2 \times 0{\cdot}2$ or 2×1 mg kg^{-1}; the last injection of dobutamine was always given at least 90 min before the experiments.

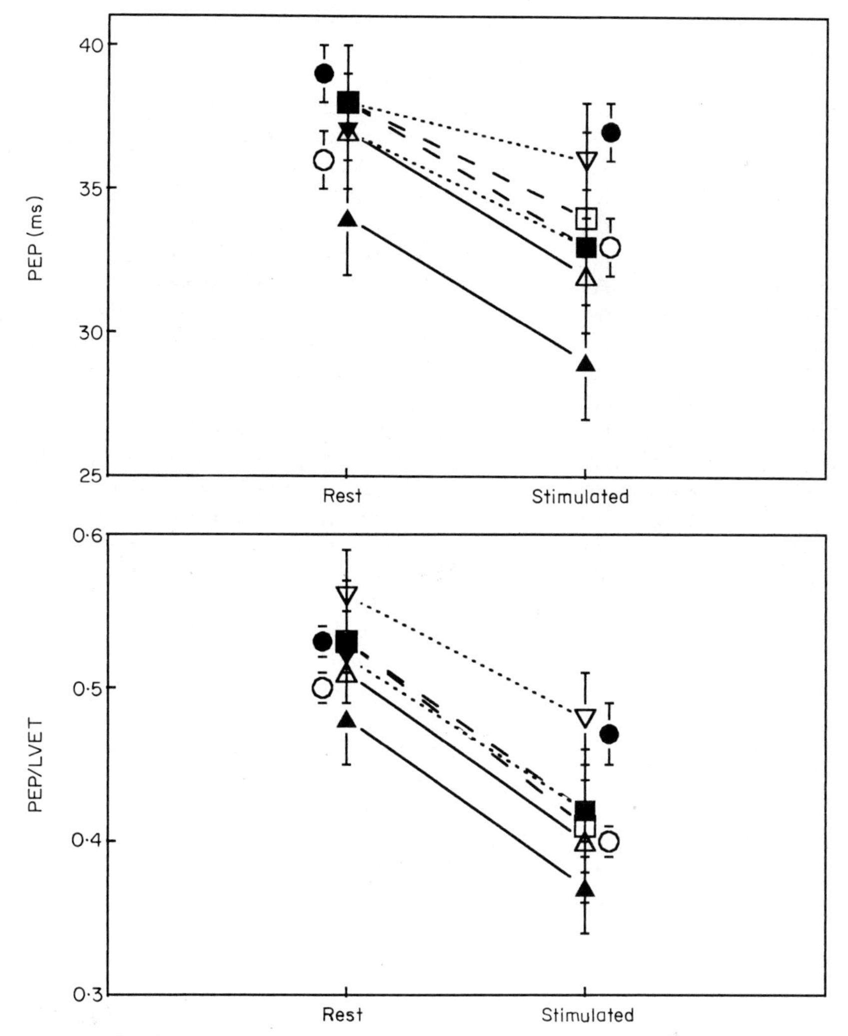

Figure 7 Resting and stimulated values for STI in sham-infarcted, infarcted, and treated infarcted rats. ○ = sham-operated animals; ● = untreated infarct animals; △, ▲ = milrinone-treated; □, ■ = dobutamine-treated; ▽, ▼ = captopril-treated. High and low doses of the drugs are represented by open and closed symbols, respectively.

Treatments had minor effects on mean arterial pressure and heart rate. Only the high dose of captopril had an effect on blood pressure. None of the drugs significantly altered HR. For dobutamine, tachycardia was absent, confirming that the drug was not present at the time of experimentation.

The highest doses of each drug improved cardiac function, improving resting and stimulated CO (Fig. 6). The associated STI, both at rest and following rapid volume loading, are summarized in Fig. 7. Volume loading decreased PEP in sham-operated animals, as expected from an intervention that increases preload and, thus, activates the Frank-Starling mechanism. Simultaneous prolongation of LVET results in a decrease of PEP/LVET. In infarcted rats, the response was similar for PEP, but the decrease of PEP/LVET was less marked. Milrinone caused dose-dependent shortening of PEP, and decrease in PEP/LVET, both at rest and following volume loading. The abbreviation of PEP is consistent with the presence of a positive inotropic substance in the circulation at the time of the experiment. Dobutamine, which was not in the circulation at the time of experiments (cf above), did not affect resting PEP or PEP/LVET. Upon stimulation, however, PEP was abbreviated dose-dependently, suggesting that the 'inotropic reserve' of the heart was increased. PEP/LVET was also decreased; however, this effect was dose-independent. Similar results were obtained for the high (effective) dose of captopril, which lacks inherent positive inotropic activity[9].

The effects of milrinone suggest that positive inotropism can be detected, both at rest and following stimulation, as a shortening of PEP, and confirm observations made in man with a variety of positive inotropes[18,31–33]. Dobutamine, once cleared from the circulation, and captopril do not have positive inotropic properties. This is reflected in a lack of effect on PEP at rest. The increased 'inotropic reserve' is, however, consistent with biochemical change following chronic ACE-inhibitor therapy[34] and dobutamine therapy[35], because both reduce the relative number of slow V_3 fibres in the rat heart.

What information can we derive from STI in rats?

In our studies in aging spontaneously hypertensive rats and in rats following myocardial infarction, PEP and PEP/LVET changed with assumed (in the SHR) or documented (in the infarcted rats) myocardial alterations. In fact, in both models these proposed indicators for the mechanical condition of the heart appear to be the most sensitive markers we can obtain. Although it is inappropriate to pool findings in these two models with different etiologies for the same disease, some similarities do exist.

PEP invariably increases in both models. In the absence of changes of QS-width, this can only indicate a decreased rate of contraction. This is consistent with both measured dP/dt in SHR[36,37] and infarcted rats[2,30] and morphological and biochemical changes in these animals[23–25,38]. Thus, in spite of the unchanged cardiac function curve in SHR and in rats with small myocardial infarcts, PEP indicates decreased rates of contraction. Although a normal cardiac function curve in a state of decreased dP/dt seems paradoxical, the cardiac function curve as we measure it not only depends upon the rate of contraction, but also upon total developed force. Hypertrophy, in both models, may compensate for the decreased rate of contraction.

Similar conclusions can be drawn from the results of long-term therapy after myocardial infarction. Improvement of cardiac function was associated with a decrease in resting PEP, in the presence of a positive inotrope (milrinone). In the absence of a positive inotropic component (dobutamine and captopril), resting PEP was not changed, but stimulated PEP was abbreviated, which we interpret as resulting from underlying biochemical changes.

In both SHR and infarcted rats we noted an increased PEP/LVET. In infarcted animals, this was accounted for by increased PEP, since LVET did not change. In contrast, SHR showed changes in LVET and LVETI, contributing to the increased ratio. Nevertheless, in neither model did we uncover a meaningful relationship between this measurement and other indicators of cardiac function, so we cannot interpret the decreases in PEP/LVET following therapy. In man, the relationship between PEP/LVET and cardiac function is purely empirical[17] and has no apparent physiological basis. From our data we must conclude that, empirically, this variable in rats does not have a physiological counterpart.

In conclusion, measurement of PEP may provide valuable information on the condition of the myocardium in rats. Depressed cardiac function curves are always associated with prolonged PEP, even at rest, whereas improved cardiac function is always associated with shortened PEP during maximal stimulation, independent of possible positive inotropic properties of the substances used to achieve improvement.

References

[1] Smith HJ, Nuttall A. Experimental models of heart failure. Cardiovasc Res 1985; 19: 181–6.

[2] Drexler H, Flaim SF, Toggart EJ, Glick MR, Zelis R. Cardiocirculatory adjustments to exercise following myocardial infarction in rats. Basic Res Cardiol 1986; 81: 350–60.

[3] Curtis JM, MacLeod BA, Walker MJA. Models for the study of arrhythmias in myocardial ischemia and infarction: the use of the rat. J Mol Cell Cardiol 1987; 19: 399–419.

[4] Pfeffer MA, Pfeffer JM, Fishbein MC *et al.* Myocardial infarct size and cardiac function in rats. Circ Res 1979; 44: 503–12.

[5] Pfeffer MA, Pfeffer JM, Frohlich ED. Pumping ability of the hypertrophying left ventricle of the spontaneously hypertensive rat. Circ Res 1976; 38: 423–9.

[6] Pfeffer J, Pfeffer M, Fletcher P, Braunwald E. Alterations of cardiac performance in rats with established spontaneous hypertension. Am J Cardiol 1979; 44: 994–8.

[7] Smith TL, Hutchins PM. Central hemodynamics in the developmental stage of spontaneous hypertension in the unanesthetized rat. Hypertension 1979; 1: 508–17.

[8] Smits JFM, Struyker-Boudier HAJ. Systemic and regional hemodynamics following acute inhibition of angiotensin I-converting enzyme in the conscious spontaneously hypertensive rat. Progr Pharmacol 1984; 5: 39–49.
[9] Schoemaker RG. Experimental heart failure in rats; hemodynamic studies on pathophysiology and therapy. The Netherlands: PhD Thesis, University of Limburg, Maastricht, 1989.
[10] Schoemaker RG, Urquhart J, Debets JJM, Struyker-Boudier HAJ, Smits JFM. Acute effects of coronary artery ligation in conscious rats. Bas Res Cardiol 1990; 85: 9–20.
[11] Schoemaker RG, Urquhart J, Debets JJM, Struyker-Boudier HAJ, Smits JFM. Cardiac function in conscious rats: acute effects of dobutamine and milrinone. Progr Pharmacol 1990; 7: 37–47.
[12] Mirsky I, Pfeffer JM, Pfeffer MA, Braunwald E. The contractile state as the major determinant in the evolution of left ventricular dysfunction in the spontaneously hypertensive rat. Circ Res 1983; 53: 767–78.
[13] Nishimura H, Kubo S, Nishioka A, Imamura K, Kawamura K, Hasegawa M. Left ventricular diastolic function of spontaneously hypertensive rats and its relationship to structural components of the left ventricle. Clin Sci 1985; 69: 571–80.
[14] Noresson E, Ricksten S-E, Thorän P. Left atrial pressure in normotensive and spontaneously hypertensive rats. Acta Physiol Scand 1979; 107: 9–12.
[15] Pfeffer JM, Pfeffer MA, Fishbein MC, Frohlich ED. Cardiac function and morphology with aging in the spontaneously hypertensive rat. Am J Physiol 1979; 237: H461–8.
[16] Schoemaker RG, Debets JJM, Halders SGEA *et al.* The aging spontaneously hypertensive rat as a model for heart failure? Submitted for publication.
[17] Lewis RP, Rittgers SE, Forester WF, Boudoulas H. A critical review of the systolic time intervals. Circulation 1977; 56: 146–58.
[18] Weissler AM, Harris WS, Schoenfeld CD. Systolic time intervals in heart failure in man. Circulation 1968; 37: 149–59.
[19] Craige E. Heart sounds. In: Braunwald E, ed. Heart disease. Textbook of cardiovascular medicine. Philadelphia: WB Saunders, 1988: 55–8.
[20] Warrington SJ, Weerasuriya K, Burgess DC. Correction of systolic time-intervals for heart rate: A comparison of individual with population derived regression equations. Br J Clin Pharm 1988; 26: 155–65.
[21] Rousson D, Galleyrand J, Silie M, Boissel JP. Uncorrected pre-injection period: a simple non-invasive measurement for pharmacodynamic screening of inotropic activity. Eur J Pharmacol 1987; 31: 559–62.
[22] Spodick DH, Doi YL, Bishop RL, Hashimoto T. Systolic time intervals reconsidered. Re-evaluation of the pre-ejection period: absence of relation to heart rate. Am J Cardiol 1984; 53: 1667–70.
[23] Vogt M, Jacob R, Kissling G, Rupp H. Chronic cardiac reactions. II. Mechanical and energetic consequences of myocardial transformation versus ventricular dilatation in the chronically pressure-loaded heart. Basic Res Cardiol 1987; 82 (Suppl 2): 147–60.
[24] Vogt M, Jacob R, Noma K, Onegi B, Rupp H. Chronic cardiac reactions. III. Factors involved in the development of structural dilatation. Basic Res Cardiol 1987; 82 (Suppl 2): 161–72.
[25] Winegrad S, McClellan G, Weisberg A, Lin LE, Weindling S, Horowits R. Ca-independent regulation of cardiac myosin. Basic Res Cardiol 1987; 82 (Suppl 2): 183–90.
[26] Garrard Jr CL, Weissler AM, Dodge HT. The relationship of alterations in systolic time intervals to ejection fraction in patients with cardiac disease. Circulation 1970; 42: 455–62.
[27] Fishbein MC, MacLean D, Maroko PR. Experimental myocardial infarction in the rat. Am J Pathol 1978; 90: 57–70.
[28] Schoemaker RG, Debets JJM, Struyker-Boudier HAJ, Smits JFM. Beneficial hemodynamic effects of 2 weeks milrinone treatment in conscious rats with heart failure. Eur J Pharm, 1990; 182: 527–35.
[29] Schoemaker RG, Debets JJM, Struyker-Boudier HAJ, Smits JFM. Delayed but not immediate captopril therapy improves cardiac function in conscious rats following myocardial infarction. Mol Cell Cardiol, 1991 (In press).
[30] Drexler H, Toggart EJ, Glick MR, Heald J, Flaim SF, Zelis R. Regional vascular adjustments during recovery from myocardial infarction in rats. J Am Coll Cardiol 1986; 8: 134–42.
[31] Svensson G, Rehnqvist N, Sjogren A, Erhardt L. Hemodynamic effects of ICI 118,587 (Corwin) in patients with mild cardiac failure after myocardial infarction. J Cardiovasc Pharm 1985; 7: 97–101.
[32] Stopfkuchen H, Schranz D, Huth R, Jungst BK. Effects of dobutamine on left ventricular performance in newborns as determined by systolic time intervals. Eur J Pediatr 1987; 146: 135–9.
[33] Lewis RP, Marsh DG, Sherman JA, Forester WF, Schaal SF. Enhanced diagnostic power of exercise testing for myocardial ischemia by addition of post-exercise left ventricular ejection time. Am J Cardiol 1977; 39: 767–75.
[34] Michel JB, Lattion AL, Salzmann JL *et al.* Hormonal and cardiac effects of converting enzyme inhibition in rat myocardial infarction. Circ Res 1988; 62: 641–50.
[35] Buttrick P, Malhotra A, Factor S, Geenen D, Scheuer J. Effects of chronic dobutamine administration on hearts of normal and hypertensive rats. Circ Res 1988; 63: 173–81.
[36] Friberg P, Nordlander M, Lundin S, Folkow B. Effects of ageing on cardiac performance and coronary flow in spontaneously hypertensive and normotensive rats. Acta Physiol Scand 1985; 125: 1–11.
[37] Friberg P, Nordborg C. Functional, morphological and metabolic characteristics of isolated hearts from normotensive and spontaneously hypertensive rats before, during and after renal hypertension. Acta Physiol Scand 1986; 126: 161–71.
[38] Geenen DL, Malhotra A, Scheuer J. Regional variation in rat cardiac myosin iso-enzymes and ATPase activity after infarction. Am J Physiol 1989; 256: H745–50.

Regulation of left ventricular pressure fall

T. C. Gillebert and D. L. Brutsaert

Department of Physiology and Department of Cardiology, University of Antwerp, Belgium

KEY WORDS: Relaxation, diastolic function, pressure fall.

Left ventricular pressure (LVP) fall is a manifestation of relaxation and is therefore regulated by non-uniformity and load, besides regulation by muscle inactivation.

Non-uniformity and load *are important and independent regulators of LVP fall. Non-uniformity induces a premature onset and decreased rate of LVP fall. With regard to load, the effects of preload, systolic LVP levels and systolic LVP waveform are distinct. In order to assess underlying muscle inactivation with indexes of LVP fall, it is critical to control non-uniformity and various aspects of load. This control should be inherent in experiments designed to evaluate interventions. If careful control of non-uniformity and load cannot be achieved, at least the effects of non-uniformity and load on timing and rate of LVP falls should be appreciated.*

Isovolumetric left ventricular pressure fall *is a complex event including a rapid early and slower late phase. As a consequence, LVP fall cannot be described by a single index. Peak −dP/dt is measured at the transition between early and late LVP fall. Its value can be affected by changes of both phases. Interpretation of peak −dP/dt should therefore be cautious. LVP fall at and after mitral valve opening might be regulated differently from late LVP fall, as evaluated by the time constant τ. Loading effects of LVP fall after mitral valve opening can therefore not always be predicted from changes in τ.*

Load dependence *is a concept based on experimental observations in isolated cardiac muscle, which describes the separation in time between shorter isotonic and longer isometric twitches. Temporal separation is illustrated by the response of cardiac muscle to late systolic load. The latter induces premature onset of relaxation, followed by accelerated isotonic, or decelerated isometric relaxation. LVP fall is regulated by load dependence. When LVP increases, LVP fall is delayed. Late LVP fall seems to be predominantly regulated by isometric load dependence, and decelerates with increased late systolic LVP. Early LVP fall can accelerate with late systolic LVP. The relation of this phenomenon to isotonic load dependence is as yet unclear.*

Introduction

Left ventricular pressure (LVP) fall is a *manifestation of relaxation*. 'Relaxation' relates to the processes whereby cardiac muscle returns, after contraction, to its initial length or tension[1]. Transition between contraction and relaxation[2] occurs after the first half of left ventricular (LV) ejection so that behaviour during later ejection is already regulated by underlying relaxation. In normal circumstances, LV relaxation is completed during rapid filling, near minimum pressure. At that moment, cardiac systole is over and diastole begins. 'Diastole' means separation between two contraction–relaxation cycles and its use should be restricted to describing passive cardiac properties such as muscular or ventricular compliance[3,4].

Address for correspondence: Dirk L. Brutsaert, MD, PhD, Department of Physiology, RUCA, University of Antwerp, Groenerborgerlaan, 171, 2020 Antwerpen, Belgium.

Analysis of LVP fall involves analysis of onset and analysis of pattern. Onset of LVP fall is related to ejection duration, systolic time intervals or time from end-diastole to peak −dP/dt. The pattern of LVP fall is complex and cannot be described by a single index. The most widely used indexes describing rate of LVP fall are peak −dP/dt, time constant of isovolumetric relaxation, τ, and isovolumetric relaxation (IR) time. Peak −dP/dt is an instantaneous value measured in the early and faster part of LVP fall. τ is a widely used mathematical description of LV pressure fall between peak −dP/dt and mitral opening, and reflects the rate of late and slower LVP fall. IR time includes both early and late pressure fall and is determined both by mean rate of LVP fall and by the pressure difference between aortic valve closure and mitral valve opening.

LVP fall is regulated by cardiac muscle inactivation[4]. Cardiac muscle inactivation has physiological determinants such as heart rate, temperature and neurohumoral stimulation. Impaired cardiac muscle inactivation can result from ageing, acute and chronic myocardial ischaemia, hypertrophy and myocardial fibrosis, hypothyreosis, myocarditis etc. Clinical cardiologists seek indexes reflecting pathological changes in cardiac muscle inactivation, which might be used in clinical situations. τ and the underlying assumption that isovolumetric LVP fall is monoexponential were initially considered[5] to be the rather undisturbed expression of cardiac muscle inactivation. However when experimental data became available, it appeared that neither τ, nor any index of relaxation, could be used to assess directly muscle inactivation. Relaxation itself, and hence its indexes, are regulated not only by muscle inactivation but also by loading conditions and by non-uniformity[2]. Therefore a distinction between contribution of impaired inactivation, non-uniformity and altered load requires detailed understanding of how LV non-uniformity and load effect the timing and pattern of LVP fall. This account intends to be descriptive, and relates in vitro muscle behaviour to in vivo LV behaviour. It does not consider basic mechanisms, which were recently reviewed[4].

Non-uniformity and left ventricular pressure fall

Mechanical LV non-uniformity can result from e.g. inhomogeneous neural stimulation, disturbances of impulse conduction, ventricular pacing, interaction between ischaemic and non-ischaemic segments or altered LV geometry. Altered geometry and mechanical non-uniformity have been described in ischaemic heart diseases, hypertrophy and cardiac overload states. A typical manifestation of mechanical non-uniformity is asynchronous early (segmental) re-extension during isovolumetric LVP fall[6]. This phenomenon is distinct from segment lengthening due to torsional shape changes and shear strain[7]. Asynchronous early re-extension of an LV wall segment implies abnormal inward motion elsewhere, in order to maintain the isovolumetric status of the LV. With acute[8,9] and chronic ischaemia[10–13], a slower rate of LVP fall has been attributed to asynchronous wall motion. However, ischaemia has multiple effects besides non-uniformity during LVP fall, e.g. alterations in load and rate of inactivation, which may be responsible for the slower rate of LVP fall. Global hypoxia, for example, slows the rate of LVP fall[14] in the absence of discrete non-uniformity. Also with ventricular pacing, the LVP fall is slower[15–17]. However, ventricular pacing eliminates the primer function of atrial contraction, induces asynchrony during segmental shortening and results in an impaired contractile response, besides non-uniformity during LVP fall. By contrast, regional inotropic stimulation provides a means to analyse the independent influence of non-uniformity on rate of LVP fall.

Regional infusion of isoproterenol (Fig. 1) shortens LV ejection duration and induces premature onset of LVP fall[18,19]. The rate of both early and late LVP fall is decreased[20,21]: peak $-dP/dt$ decreases (is less negative); τ increases; and IR time accordingly increases. The effects of regional infusion of isoproterenol on the extent of asynchronous wall movements and rate of LVP fall are similar over a wide range of loading conditions[21]: loading conditions therefore interact neither with extent nor with effects of non-uniformity.

Non-uniformity is an important modulator of the LVP fall, which exerts its influence by mechanisms independent of loading conditions.

Load and left ventricular pressure fall

The effects of diastolic load, systolic load and systolic load pattern are considered here both in vitro (isolated feline cardiac muscle) and in vivo (intact ejecting canine LV). The isometric force decline of isolated muscle is analysed as a model for isovolumetric pressure fall of the LV. Data are then integrated by describing the effects of a steady-state increase in LVP.

PRELOAD AND DIASTOLIC PRESSURES

In isolated cardiac muscle, preload changes induce subtle but consistent changes in timing and rate of isometric force decline[22]. Acute preload reduction results in premature onset and increased rate of isometric force decline. When preload reduction is maintained at a reduced level, the decline in force is progressively delayed and slowed. Accordingly, chronic preload reduction does not clearly affect timing or rate of force decline[23,24] because distinct and opposite influences on relaxation cancel each other[22].

In the intact LV, acute or chronic increases in diastolic LVP (with no change in systolic LVP) slightly prolong LV ejection[23]. No effects of diastolic LVP on peak $-dP/dt$ or τ could be demonstrated[23,26]. Diastolic LVP, therefore, affects timing, but not rate of LVP fall.

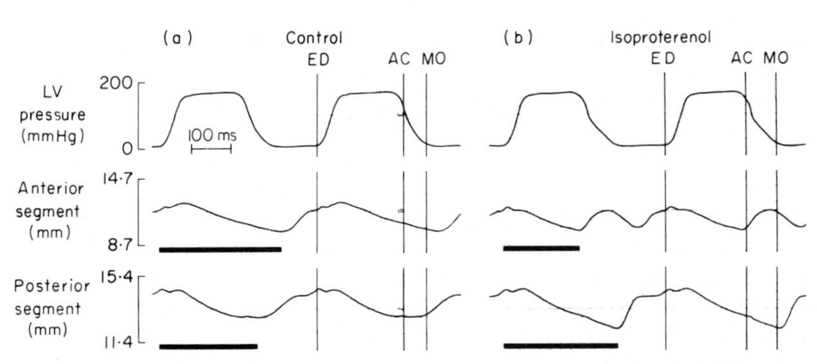

Figure 1 Asynchronous wall movements and left ventricular pressure (LVP) fall. Tracings of LVP, anterior and posterior segment lengths. Vertical lines denote timing of end-diastolic (ED), aortic valve closure (AC), and mitral valve opening (MO). During control period (Panel a), both segments shorten synchronously and show minimal change in length during isovolumetric LVP fall (AC to MO). After injection of 10 ng of isoproterenol into the mid-left anterior descending coronary artery (Panel b), the anterior segment has increased rate of shortening, followed by marked premature lengthening during LVP fall. The posterior segment is not directly stimulated but shows increased shortening during LVP fall. There is a regional intraventricular unloading effect whereby the posterior wall shortens and shifts blood into the expanding anterior wall during LVP fall. This increase in non-uniformity is associated with an earlier onset (shorter interval ED–AC) and slower overall rate (longer interval AC–MO) of LVP fall.

SYSTOLIC LOAD LEVELS AND PRESSURES

In isolated cardiac muscle, increasing afterload delays the onset of force decline[4,22]. This delay is the manifestation of load dependence and separation in time between isotonic and isometric twitches[4]. Increasing afterload accelerates the force decline, increasing both peak rate of early[26] and rate of late force decline[22]. These findings in isolated muscle cannot easily be extrapolated to the intact LV.

An intervention which is similar to a selective increase in afterload of isolated muscle is an abrupt beat-to-beat increase in systolic LVP. Abrupt beat-to-beat increases in systolic LVP, with no changes in long-term load history, no changes in diastolic LVP and no changes in systolic LVP waveform (early LVP increases) can be induced by partial aortic occlusion. Such interventions are limited to pressure increments of ± 20 mmHg. Larger increments inevitably distort the systolic LVP waveform, increasing course and inducing late peaking[27]. Early LVP increases prolong LV ejection[27–29], illustrating that the in vivo situation is load dependent. Moderate ($\leqslant 20$ mmHg) early LVP increases do not significantly affect the rate of LVP fall[27]; peak $-dP/dt$ tends to increase slightly (acceleration of early LVP fall) and τ tends to be prolonged slightly (deceleration of late LVP fall). IR time increases, mainly as a consequence of a greater extent rather than slower rate of LVP fall.

The upper panels of Fig. 4 illustrate early LVP increase. The phase-plane analysis (right panel) shows the effects on the pattern of LVP fall, which can be divided into three phases. Phase I is initial acceleration (on the right of the panel). Phase II is the medium part following peak $-dP/dt$, from which τ is calculated. Phase III is late LV pressure fall (on the left of the panel) at and after mitral valve opening. During phase I, LVP fall accelerates or $-dP/dt$ increases. The control and test tracings are separated: the latter appearing to the right of the former, toward higher systolic LVP. The two curves are not parallel, but converge: the acceleration in the test tracing is, in this example, less pronounced than that in the control tracing. Peak $-dP/dt$ is similar, despite a difference in peak LVP. During phase II following peak $-dP/dt$, the two curves are dissociated and the test tracing appears slightly above the control tracing. This reflects a small increase in τ. During phase III on the left of the panel, the two tracings coincide.

SYSTOLIC LOAD PATTERN AND PRESSURE WAVEFORM

LVP waveform during ejection (systolic pressure waveform) is determined by the interaction of the ejecting left ventricle with aortic input impedance, and is modified by arterial pressure reflections[30,31]. Alterations in aortic input impedance or arterial wave reflections occur under physiologic

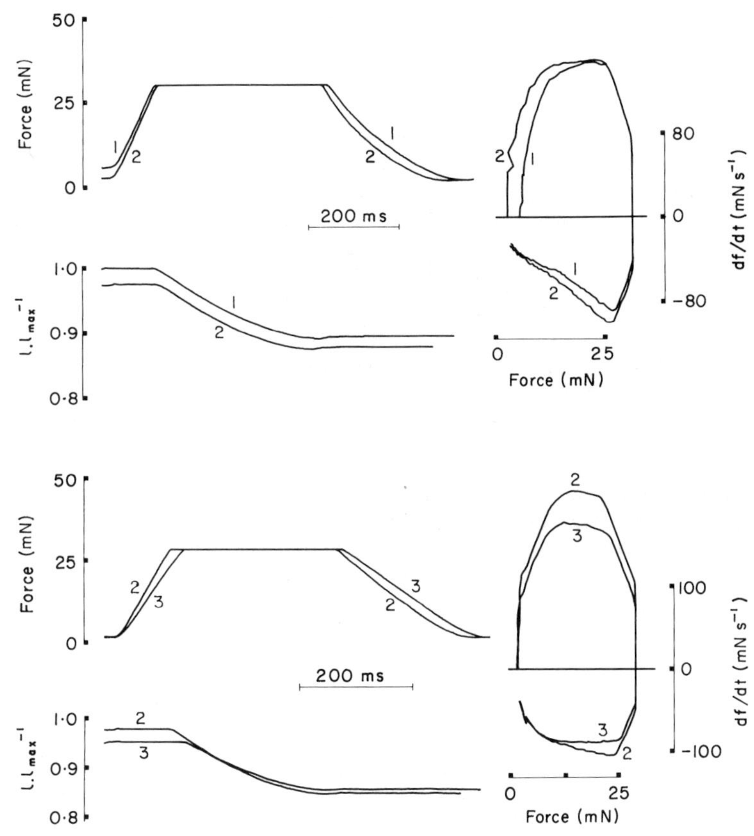

Figure 2 Preload and force decline. Force-time tracings (upper left panel), length-time tracings (lower left panel) and df/dt (rate of force change) vs f (force) tracings or phase-plane tracings (right panel). Length-time tracing is not shown beyond the isometric force decline because delay and damp functions were maintained so we did not correct for transition artifacts at the onset of isotonic lengthening. Two twitches are superposed in each panel.

Upper panels (acute preload reduction): twitch 1 is a control twitch starting from L_{max}. Twitch 2 had its preload reduced to 50% of the initial value shortly before the twitch. Both twitches have the same total systolic load. Force decline of twitch 2 starts earlier. In the phase-plane tracing, the force decline of twitch 2 is initially beneath the force decline of twitch 1, indicating a faster initial force decline. The two tracings coincide during the late force decline.

Lower panels (chronic vs acute preload reduction): twitch 2 had its preload reduced to 50% of the initial value before the twitch, as in the upper panel. Twitch 3 had preload reduced for more than 20 min, and shows delayed onset of force decline. The phase-plane tracing of twitch 3 during force decline is initially above the tracing of twitch 2, indicating a slower initial force decline. The two tracings coincide during the late force decline.

and pathologic conditions. During the Valsalva manoeuvre, for example, the systolic pressure waveform can be modified from an increasing course and late peaking, to a decreasing course and early peaking[31]. The effects of systolic pressure waveform on LVP fall can be analysed by examining the effects of acute increases in load (pressure) at different moments during systole.

The responses of isolated cardiac muscle to late systolic load are different from those to early systolic load. The former further illustrate load dependence and a separation in time between

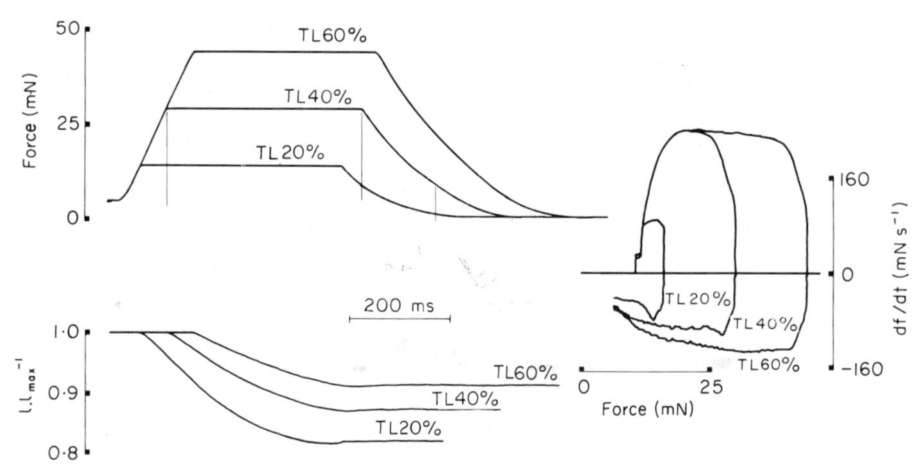

Figure 3 Afterload and force decline. Three twitches with different afterload are superposed. When afterload increases and extent of shortening decreases, the onset of force decline is markedly delayed. Phase-plane tracings indicate that, when afterload increases, force decline (tracings below baseline) is faster over the entire range. TL = total load (preload + afterload) expressed as % of peak.

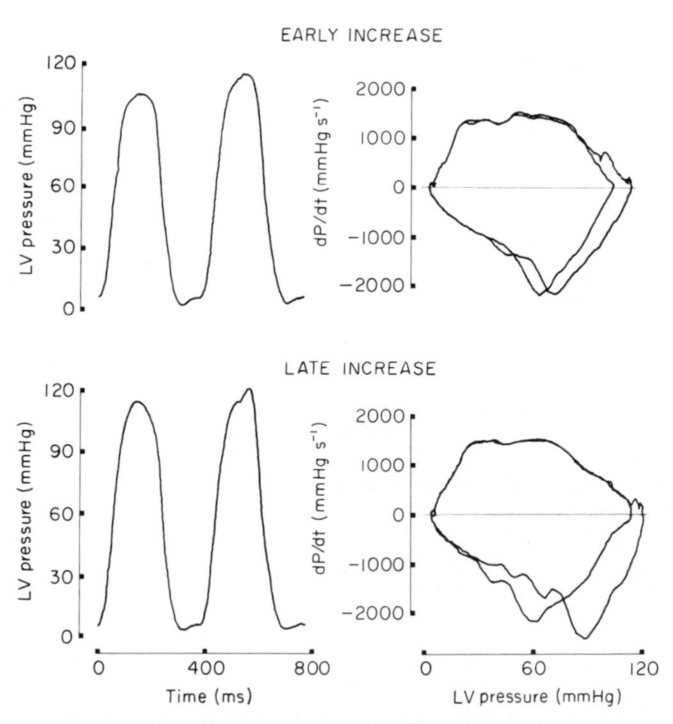

Figure 4 Systolic LVP increase and LVP fall. Pressure-time tracing is displayed on the left, phase-plane tracing (dP/dt vs LV pressure) on the right. Two consecutive heart cycles are represented: a baseline cycle and a cycle during which systolic LVP is increased by inflation of an aortic balloon. This increase occurs early during ejection in the upper panels and later during ejection in the lower panels. See text for details.

isometric and isotonic twitches. A late systolic increase in load induces early onset of relaxation, while a late systolic decrease in load delays the onset of relaxation[32]. The isotonic response (isotonic–isometric relaxation sequence) to late systolic load is an accelerated muscle lengthening and subsequent force decline[32]. The isometric response to late systolic load (isometric-isotonic relaxation sequence) is a slower force decline, preceding muscle lengthening[33]. Accordingly, the isometric response to late systolic unloading is a delayed, but faster, force decline[33].

In the intact ventricle, LVP fall is isovolumetric and occurs at minimum LV volume. Similar to the force decline of a muscle twitch with an isometric–isotonic relaxation sequence, a load-dependent response to late systolic LVP increase will manifest predominantly as deceleration of LVP fall. When systolic LVP increases occur during the second half of LV ejection, the onset of LVP fall is premature: ejection duration is abbreviated and time to peak $-dP/dt$ decreases[27–29]. Peak $-dP/dt$ can increase[29]: in the intact ejecting left ventricle, increased peak dP/dt is only observed following abrupt and rather late systolic LVP increases, as shown, for example, in Fig. 4 (lower panels). The physiological significance of this phenomenon is questionable. When pressure increases have a more gradual physiological course, peak $-dP/dt$ does not change[27]. In the intact ejecting LV, τ is markedly prolonged, reflecting a slower late LVP fall[27,34]. IR time is increased, reflecting both longer and slower LVP fall[27]. These findings suggest that alteration of the systolic pressure waveform is a mechanism influencing τ that is independent of the effects of increased systolic pressures.

The lower panels of Fig. 4 illustrate late systolic LVP increase. During phase I, or early LVP fall, $-dP/dt$ accelerates. The two curves are separated and diverge: acceleration in the test tracing is more marked than that in the control tracing. As a consequence, peak $-dP/dt$ in the test tracing is increased (more negative). During phase II, following peak $-dP/dt$, the two curves are clearly dissociated, the test tracing appearing above the control tracing, which reflects an increase in τ. During phase III, on the left of the panel, the two tracings coincide. Peak $-dP/dt$ is a transition point and its actual value depends on both phase I and phase II. In this example the early LVP fall is accelerated and the late LVP fall is slowed. Peak $-dP/dt$ is increased (more negative) since the former effect predominates.

LVP fall therefore responds differently to late than to early LVP increases. This modified responsiveness illustrates the effects on LVP fall of the LVP waveform during ejection. This behaviour is of particular importance when the systolic load pattern is abnormal, as occurs, for example, in chronic aortic regurgitation[35–37] or as in acute[38] and sometimes chronic[36,37,39] mitral regurgitation. The systolic load pattern in aortic regurgitation typically has an upward course and shows late peaking, which ought to decelerate LVP fall. The systolic load pattern in mitral regurgitation typically shows early peaking and a downward course, which ought to accelerate LVP fall.

Phase-plane analysis and subdivision of LVP fall into three separate phases show the particular position of peak $-dP/dt$, at the transition between phase I and II. Moreover, directional changes in LVP fall after mitral valve opening (phase III), with abrupt LVP increases, cannot be predicted from changes in τ. Hence, the regulation of LVP fall after mitral valve opening is presumably different from the regulation of τ.

STEADY-STATE INCREASES IN SYSTOLIC PRESSURES

The effects of beat-to-beat LVP increases should be considered as distinct from the more complicated effects of steady-state increases in LVP, as produced by methoxamine, phenylephrine, aortic occlusion or volume loading[26,34,40,41]. In these steady-state load increases, peak $-dP/dt$ increases[5,42], indicating a faster early LVP fall, and τ is prolonged, reflecting a slower late LVP fall[26,34,40,41]. Different responses to beat-to-beat and steady-state increases in load may be related to a different magnitude of intervention, to altered systolic LVP waveform, to altered neurohumoral stimulation, or to volume-dependent changes in contractility[43].

A typical example of a steady-state increase in load is volume loading. The effects of volume on LVP fall have been extensively analysed[26,40,41] previously and are illustrated by data recently obtained in anaesthetized open-chest dogs ($n = 7$)[44].

In Fig. 5, three LVP curves are superposed. When end-diastolic pressure (EDP) increases from low to high levels, systolic LVP increases and develops an upward course with late peaking. LVP fall is delayed. This delay results both from load dependence[4] and from a length-dependent increase in contractility[43].

Table 1 reveals that the early LVP fall accelerates (peak $-dP/dt$ increases) and that late LVP fall decelerates (τ increases). In this series, IR time does

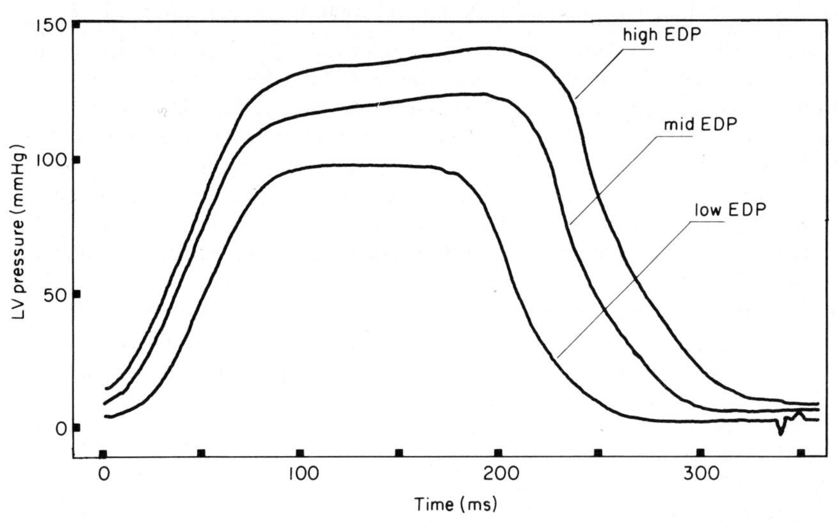

Figure 5 Volume loading and LVP fall. Three LVP tracings are superposed corresponding to three steady-state conditions. EDP is end-diastolic LVP. When EDP increases from low to high levels, systolic LVP increases. The systolic LVP waveform is modified, developing an upward course and late peaking. The LVP fall is delayed. See Table 1 for an analysis of LVP fall.

Table 1 Effects of volume loading on LVP fall

	Low EDP	Mid EDP	High EDP	*P* value
LVEDP (mmHg)	4·1 ± 0·9	10·6 ± 1·5	17·9 ± 1·8	<0·001
Systolic LVP (mmHg)	111 ± 16	146 ± 17	178 ± 35	<0·001
Time to −dP/dt (ms)	188 ± 23	212 ± 22	222 ± 29	<0·001
Peak −dP/dt (mmHg s^{-1})	2570 ± 579	3649 ± 637	3983 ± 641	<0·001
τ (ms)	24 ± 5	28 ± 5	34 ± 18	<0·001
IR time (ms)	59 ± 16	53 ± 18	54 ± 17	N.S.

not vary from low to high levels of EDP as the effects of faster early LVP fall are cancelled by the effects of a slower late LVP fall. In previous studies, the slowing effect of volume loading predominated, and IR time increased[21]. At present it is difficult to interpret the effects of volume loading on the basis of isolated cardiac muscle findings. In volume loading, non-uniformity does not change[21]. Acceleration of early LVP fall appears to be similar to the acceleration of force decline with increasing afterload, and deceleration of the late LVP fall is related to an altered systolic LVP waveform. The extent that early segmental re-extension, due to torsional movements, and shear strain[7] also contribute to the slower late LVP fall is still unknown.

References

[1] Hill AV. The energetics of relaxation in a muscle twitch. Proc Roy Soc London B, Biol Sci 1949; 136: 211–9.
[2] Brutsaert DL, Rademakers FL, Sys SU, Gillebert TC, Housmans PR. Analysis of relaxation in the evaluation of ventricular function of the heart. Prog Cardiovasc Dis 1985; 28: 143–63.
[3] Aubert H. Innervation der Kreislauforgane. 1880.
[4] Brutsaert DL, Sys SU. Relaxation and diastole of the heart. Physiol Rev 1989; 69: 1228–315.
[5] Weiss JL, Frederiksen JW, Weisfeldt ML. Hemodynamic determinants of the time course of fall in canine left ventricular pressure. J Clin Invest 1976; 58: 751–60.
[6] Gaasch WH, Blaustein AS, Bing OHL. Asynchronous (segmental early) relaxation of the left ventricle. J Am Coll Cardiol 1985; 5: 891–7.
[7] Waldman LK, Fung YC, Covell JW. Transmural myocardial deformation in the canine left ventricle. Circ Res 1985; 57: 152–63.

[8] Kumada T, Karliner JS, Pouleur H, Gallagher KP, Shirato K, Ross JJr. Effects of coronary occlusion on early ventricular diastolic events in conscious dogs. Am J Physiol 1979; 237: H542–9.

[9] Waters DD, Da Luz P, Wyatt HL, Swan HJC, Forrester JC. Early changes in regional and global left ventricular function induced by graded reductions in regional coronary perfusion. Am J Cardiol 1977; 39: 537–43.

[10] Carroll JD, Hess OM, Hirzel HO, Krayenbuehl HP. Exercise-induced ischemia: the influence of altered relaxation on early diastolic pressures. Circulation 1983; 67: 521–8.

[11] Papapietro SE, Coghlan HC, Zissermann D, Russel RO, Rackley CE, Rodger WJ. Impaired maximal rate of left ventricular relaxation in patients with coronary artery disease and left ventricular dysfunction. Circulation 1979; 59: 984–91.

[12] Rousseau MF, Veriter C, Detry JR, Brasseur L, Pouleur H. Impaired early left ventricular relaxation in coronary artery disease: effects of intracoronary nifedipine. Circulation 1980; 62: 764–72.

[13] Ludbrook PA, Byrne JD, Tiefenbrunn AJ. Association of asynchronous protodiastolic segmental wall motion with impaired left ventricular relaxation. Circulation 1981; 64: 1201–11.

[14] Palacios I, Newell JB, Powell WJ Jr. Effects of acute global ischemia on diastolic relaxation in canine hearts. Am J Physiol 1978; 235: H720–7.

[15] Blaustein AS, Gaasch WH. Myocardial relaxation VI: effects of beta-adrenergic tone asynchrony on LVP and relaxation rate. Am J Physiol 1983; 244: H417–22.

[16] Heyndrickx GR, Van Trimpont PJ, Rousseau MF, Pouleur H. Effects of asynchrony on myocardial relaxation at rest and during exercise in conscious dogs. Am J Physiol 1988; 254: H817–22.

[17] Zile MR, Blaustein AS, Shimizu G, Gaasch WH. Right ventricular pacing reduces the rate of left ventricular relaxation and filling. J Am Cardiol 1987; 10: 702–9.

[18] Ilebekk AJ, Lekven J, Kiil F. Left ventricular asynergy during intracoronary isoproterenol infusion in dogs. Am J Physiol 1980; 239: H594–600.

[19] Gwirtz PA, Franklin D, Mass HJ. Modulation of synchrony of left ventricular contraction by regional adrenergic stimulation in conscious dogs. Am J Physiol 1986; 251: H490–5.

[20] Lew WYW, Rasmussen CW. Influence of nonuniformity on rate of left ventricular pressure fall in the dog. Am J Physiol 1989; 257: H222–32.

[21] Gillebert TC, Lew WYW. Nonuniformity and volume loading independently influence isovolumic relaxation rates. Am J Physiol 1989; 257: H1927–35.

[22] Gillebert TC, Raes DF. Preload and isometric force decline. (in preparation).

[23] Gaasch WH, Carroll JD, Blaustein AS, Bing OHL. Myocardial relaxation: effects of preload on the time course of isovolumetric relaxation. Circulation 1986; 73: 1037–41.

[24] Zile MR, Conrad CH, Gaasch WH, Robinson KG, Bing OHL. Preload does not affect relaxation rate in normal, hypoxic or hypertrophic myocardium. Am J Physiol 1990; 258: H191–7.

[25] Wiegner AW, Bing OHL. Isometric relaxation of rat myocardium at end-systolic fiber length. Circ Res 1978; 43: 865–9.

[26] Gaasch WH, Blaustein AS, Andrias CW, Donahue RP, Avitall B. Myocardial relaxation II: Hemodynamic determinants of rate of LV isovolumic pressure decline. Am J Physiol 1980; 239: H1–6.

[27] Gillebert TC, Lew WYW. Timing of abrupt systolic pressure increases and the rate of left ventricular pressure fall. Am J Physiol (submitted for publication).

[28] Noble MIM. The contribution of blood momentum to left ventricular ejection in the dog. Circ Res 1968; 23: 663–70.

[29] Ariel Y, Gaasch WH, Bogen DK, McMahon TA. Load-dependent relaxation with late systolic volume steps: servo-pump studies in the intact canine heart. Circulation 1987; 75: 1287–94.

[30] Elzinga G, Westerhof N. Pressure and flow generated by the left ventricle against different impedances. Circ Res 1973; 32: 178–86.

[31] Murgo JP, Westerhof N. Arterial reflections and pressure waveforms in humans. In: Yin FCP ed. Ventricular vascular coupling: Clinical, physiological and engineering aspects. New York: Axel Springer Verlag 1987: 140–58.

[32] Brutsaert DL, De Clerck NM, Housmans PR, Goethais MA. Relaxation of ventricular cardiac muscle. J Physiol (London) 1978; 283: 469–80.

[33] Gillebert TC, Sys SU, Brutsaert DL. Influence of loading patterns on peak length-tension relation and on relaxation in cardiac muscle. J Am Coll Cardiol 1989; 13: 483–90.

[34] Hori M, Inoue M, Kitakaze M *et al.* Loading sequence is a major determinant of afterload-dependent relaxation in intact canine heart. Am J Physiol 1985; 249: H747–54.

[35] Osbakken M, Bove AA, Spann JF. Left ventricular function in chronic aortic regurgitation with reference to end-systolic pressure, volume and stress relations. Am J Cardiol 1981; 47: 193–8.

[36] Wisenbaugh T, Spann JF, Carabello BA. Differences in myocardial performance and load between patients with similar amounts of chronic aortic versus chronic mitral regurgitation. J Am Coll Cardiol 1984; 3: 916–23.

[37] Gillebert TC, Rademakers FE, Brutsaert DL. La fonction du ventricule gauche dans les valvulopathies acquises: analyse de la relaxation. In: Acar J, ed. Les cardiopathies valvulaires acquises. Paris: Flammarion Medecine-Science, 1985: 188–200.

[38] Urschel CW, Covell JW, Sonnenblick EH, Ross JJr, Braunwald E. Myocardial mechanics in aortic and mitral regurgitation: the concept of instantaneous impedance as a determinant of the performance of the intact heart. J Clin Invest 1968; 47: 867–83.

[39] Corin WJ, Scott Monrad E, Murakami T, Nonogi H, Krayenbuehl HP. The relationship of afterload to ejection performance in chronic mitral regurgitation. Circulation 1987; 76: 59–67.

[40] Karliner JS, LeWinter MM, Mahler F, Engler R, O'Rourke RA. Pharmacologic and hemodynamic influences on the rate of isovolumic left ventricular relaxation in the normal conscious dog. J Clin Invest 1977; 60: 511–21.

[41] Raff GL, Glantz SA. Volume loading slows left ventricular isovolumic relaxation rate. Evidence of load-dependent relaxation in the intact dog heart. Circ Res 1981; 48: 813–24.

[42] Weisfeldt ML, Scully HE, Frederiksen J *et al.* Hemodynamic determinants of maximum $-dP/dt$ and periods of diastole. Am J Physiol 1974; 227: 613–21.
[43] Lew WYW. Time-dependent increase in left ventricular contractility following acute volume loading in the dog. Circ Res 1988; 63: 635–47.
[44] De Hert SG, Gillebert TC, Jagenau AH, Brutsaert DL. Endocardial modulation of left ventricular performance depends on prevailing load. Circulation 1990; 82 (Suppl III): 567.

Potential application of new imaging modalities in the measurement of cardiac volumes and characteristics

W. J. MacIntyre

Department of Nuclear Medicine, Division of Radiology, Cleveland Clinic Foundation, Cleveland, Ohio, U.S.A.

KEY WORDS: Positron emission tomography, magnetic resonance imaging, cine-computed tomography, ventricular volumes, myocardial thickness.

Developments of various imaging modalities in recent years have resulted in imaging systems with greater spatial resolution, faster data acquisition, and more powerful computer processing. It is the purpose of this report to examine three of these systems, positron emission tomography (PET), magnetic resonance imaging (MRI) and ultrafast computed tomography (cine-CT) with respect to their applicability for determination of cardiac volumes and other characteristics related to hypertension.

While each of these modalities has widespread applications and specific advantages, this report will be primarily concerned with two objectives: functional ability to record indicator dilution curves suitable for quantification of flow and the anatomic or geometric ability to measure chamber dimensions with high accuracy.

Time concentration curves have been outlined on all modalities and used for quantification of myocardial perfusion and myocardial reserve. Only one modality, cine-CT, determined cardiac output by dilution curves although in vivo arterial recording from PET has demonstrated suitable curves for this application.

Both MRI and cine-CT have demonstrated accurate determination of chamber volumes by the geometric method, the latter giving somewhat higher spatial resolution. The advantages of these machines for such purposes have yet to be determined.

Introduction

Since the first commercial computed tomography system was announced in 1972[1], there has been a steady progression of advances in imaging technology. Common to all of these modalities have been computer analysis and display which have resulted in images which are similar in appearance but quite different in origin. Data acquisition of these modalities has differed greatly and has involved detection of a wide range of energy forms such as radiofrequency waves, sound waves, X-rays, positron decay and gamma rays.

This variance has resulted in distinct advantages and limitations for each imaging modality, which in turn has directed certain modalities to specific types of physiological or clinical applications. It is the purpose of this report to examine three of the new imaging systems, positron emission tomography (PET), magnetic resonance imaging (MRI), and cine-computed tomography (cine-CT) with respect to measurement of cardiac volumes and characteristics. This report will not attempt to review the characteristics of each imaging system or to make a comprehensive evaluation of the many measurements already performed on them. Instead, it will try to project the possible involvement or advantage these systems might have compared with existing methodology for the measurement of cardiac volumes, cardiac output and various characteristics of hypertension.

All of these modalities can theoretically visualize the individual cardiac chambers so that a geometric calculation of chamber size at systole and diastole is feasible and chamber volumes, ejection fractions, stroke volume and cardiac ouput can all be derived. In addition, these systems share a technique in common with the conventional scintillation camera, that a dilution curve from a specific site may be obtained from the passage of a bolus injection of radioactive or contrast material.

Since the various imaging systems exhibit different inherent qualities that will determine their

Address for correspondence: Dr William J. MacIntyre, Department of Nuclear Medicine, Gb3, Cleveland Clinic Foundation, 9500 Euclid Avenue, Cleveland, Ohio, U.S.A. 44195-5074.

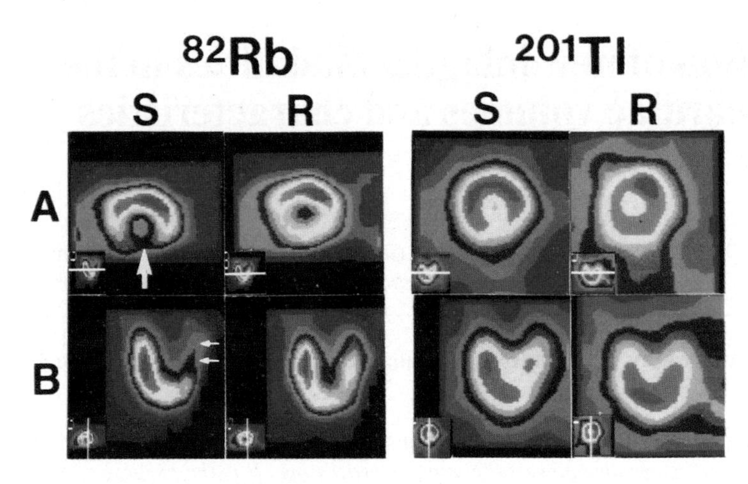

Figure 1 Comparison of ^{82}Rb PET and ^{201}Tl SPECT images from a patient with an inferior wall transient defect (ischaemia). Row A shows cardiac short axis images with anterior wall at top and septum on the left. Row B is vertical long axis with anterior wall on left and posterior wall on right. S denotes stress and R denotes rest or redistribution. Note the reduced spatial and contrast resolution on the ^{201}Tl images. Reproduced from the *Journal of Nuclear Medicine*[2], with permission.

potential applications, these systems will be first considered separately.

Positron emission tomography (PET)

There are only a few differences between PET imaging and conventional emission tomography with single photon emitters (SPECT), but these differences are very important. Positron decay results in the production of two 511 keV photons 180° apart, and the coincident recording of these photons results in a higher spatial resolution than can usually be obtained by collimators alone, such as those used with single photon emitters. In addition to increased spatial resolution, PET units may also demonstrate increased contrast resolution so that smaller differences of tracer concentration can be identified. The effects of these characteristics are demonstrated in Fig. 1 which shows sections of a heart displayed in the short and long axes with both PET and SPECT techniques[2]. The PET image appears superior in resolution but in many cases the diagnosis of ischaemia or scar may be similar. Note the differences in both the spatial and contrast resolution.

Geometric techniques applied to calculate chamber size, stroke volume, or wall thickness are more adaptable to PET images than to SPECT images, but these still fall short of the resolution necessary for satisfactory results. For most PET units inherent spatial resolution of 5 to 6 mm can be attained but normal reconstruction and processing usually broaden this resolution to 12 mm or more.

The studies shown in Fig. 1 are ungated images. Some improvement in resolution can be expected to be achieved with gating which to date has not been routinely applied to PET imaging, but is to be expected that this will be achieved with the newer PET units.

The more important advantage of the PET technique is in functional imaging made possible by such positron emitters as ^{15}O, ^{11}C, ^{13}N, and ^{18}F which can be easily incorporated into various metabolic tracers. An example of functional images is demonstrated in Fig. 2 showing short axis images of the left ventricle with perfusion images with ^{82}Rb and metabolic images of the same section with ^{18}F deoxyglucose (FDG). The defect in the resting perfusion image shows a segment not perfused and thus indicative of a scar. The increased uptake of FDG in that region, however shows the segment to be viable, and it may be considered as hibernating myocardium[3].

Dilution curves from various sites of the myocardium or chambers can be derived either by fast framing rates or by list mode acquisition. Arterial input curves have been recorded utilizing 4 s acquisitions[4] and have correlated well with samples

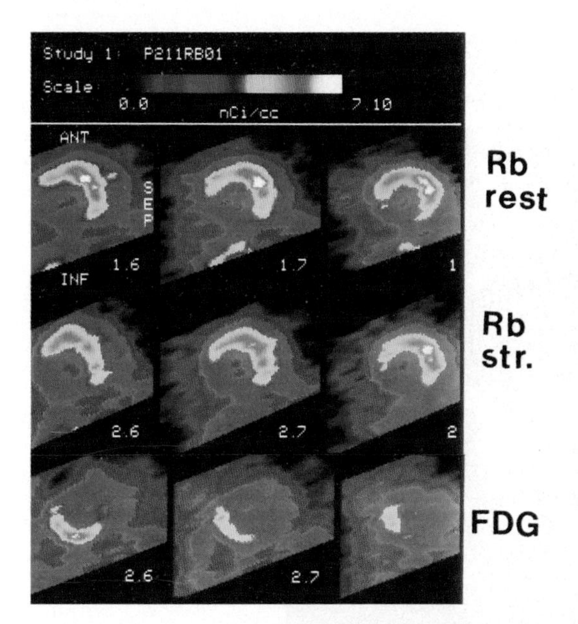

Figure 2 Top row shows ^{82}Rb perfusion image at rest, second row shows ^{82}Rb perfusion image with dipyridamole stress and bottom row shows ^{18}FDG metabolic image following exercise. Note the persistent decrease seen as a defect in the stress ^{82}Rb is the same location as a hot spot in the ^{18}FDG image.

withdrawn from the left ventricle. These curves could, of course, be used to calculate cardiac output if desired.

Magnetic resonance imaging (MRI)

In magnetic resonance imaging a considerable improvement in spatial resolution is noted over that of emission tomography. In MRI spatial localization of the radiofrequency signal is achieved by analysis of the frequency and phase of the signal originating from the object following stimulation from radiofrequency waves and magnetic gradients. No collimation is required and the pixel size can be reduced to a dimension as small as desired, noting that as the pixel size decreases, the intensity of the signal falls markedly.

The reduction of signal by collimation in PET, either electronic or by attenuation apertures, is more limiting to signal intensity than in MRI since the radioactive decay process entails a random fluctuation of emission many fold less than either X-ray or radiofrequency flux. Thus while the pixel size of PET images is usually in the 1·7 mm range the inherent resolution may be no smaller than 5·0 mm. To reduce the statistical fluctuations temporal or spatial filters may be added which broaden the inherent resolution two or threefold. This type of smoothing is not required in MRI so the inherent spatial resolution may be preserved.

The pixel sizes of magnetic resonance images are usually not much smaller than those of emission tomography, ranging from 1·25 mm to 1·56 mm. The spatial dependence of magnetic field encoding is sufficiently accurate, however, that these spatial dimensions can be maintained so that not only can the heart configuration be accurately delineated, but actual valves can be visualized. These studies[5] have used cardiac gating so a sequence of frames can be recorded continuously through a cardiac cycle, or images may be recorded at specific points of the cycle, i.e. at diastole and systole.

A second factor that has been important from early investigations of cardiac imaging with MRI is that moving blood exhibits a signal intensity different from that of the myocardium. Initial studies with the spin echo technique recorded moving blood at a lower intensity, which gave a convenient differentiation of endocardium from moving blood and a delineation of chamber dimensions with high accuracy readily achievable.

A nine segment time series of 50 ms increments from a midventricular slice[5] is illustrated in Fig. 3. The region of the descending aorta shows a high intensity signal arising from stationary blood in the aorta in the first two frames at diastole with the intensity decreasing with the increasing velocity as systole is reached.

A series of frames taken in short axis projection in both systole and diastole have been used to produce contours drawn along the endocardium[6], shown as a stack in Fig. 4a. The polygonal surface drawn from these systolic contours is shown in Fig. 4b. The difference from the volume of the polygons at diastole and systole gives the stroke volume, and the ratio of this volume to the diastole volume would yield the ejection fraction.

The ratio of the left to right ventricular stroke volume of six patients reported by investigators at Massachusetts General Hospital[6] showed a correlation coefficient of 0·95 with standard error of 4·2 ml. A somewhat later technique[7] has utilized flip angles of 30°, 12 ms gradient-refocused echoes and a repetition time of 21 ms. These measurements allow the entire heart to be visualized in 10 mm thick slices and 16 to 32 time frames in the range of 15–31 min. Twelve time frames for a slice at the midventricular level are shown in Fig. 5.

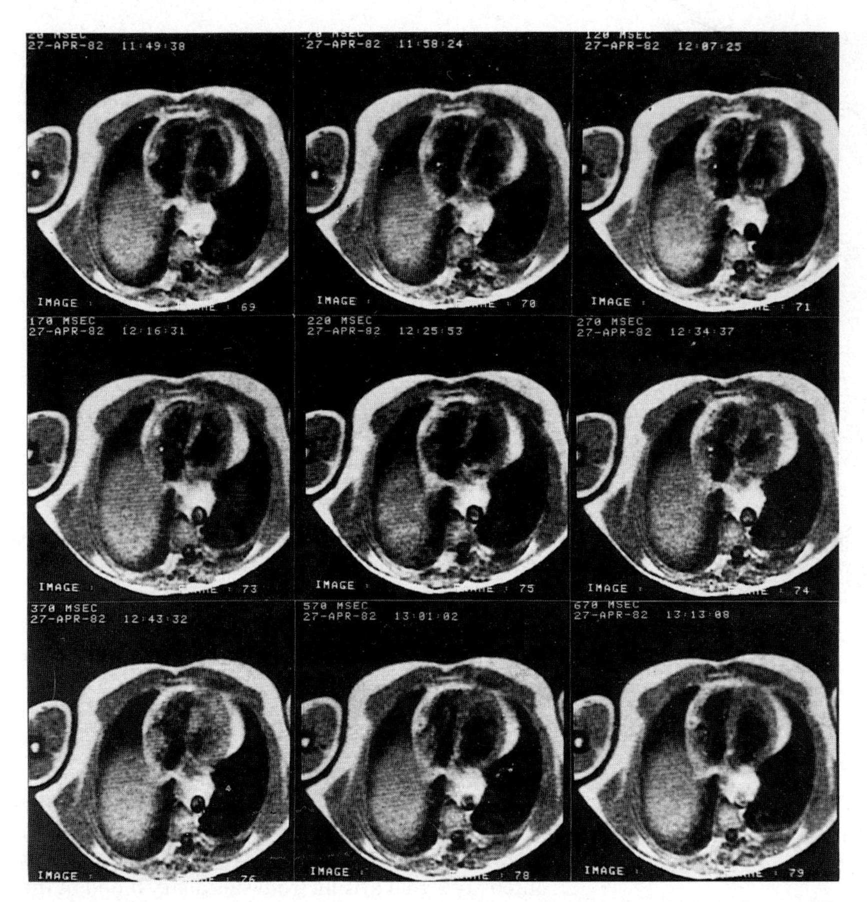

Figure 3 Time sequence at 50 ms increments of a midventricular section of the heart starting 20 ms after the R wave at upper left and progressing to 670 ms at lower right. Note the variation of intensity of the decending aorta, with a strong signal at diastole and a weak signal at systole. From George CR, Jacobs G, MacIntyre WJ *et al. Radiology* 1984; 151: 421–8, reproduced with permission.

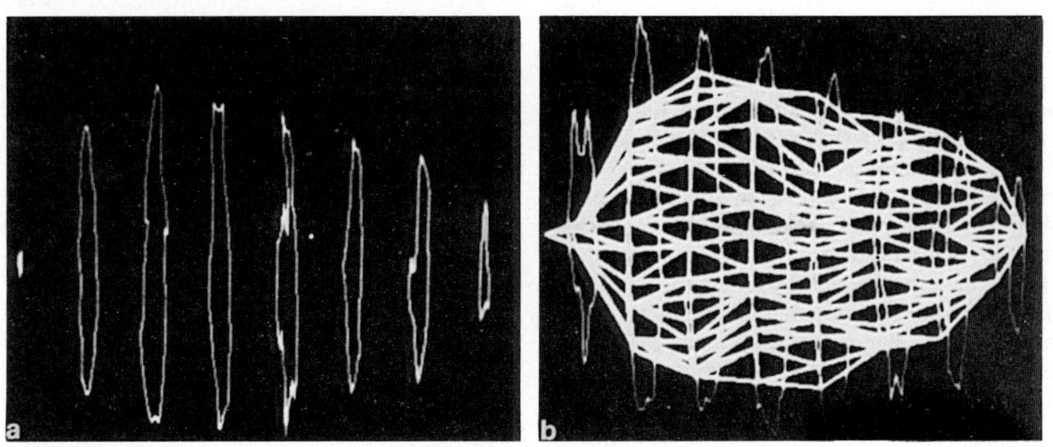

Figure 4 (a) Stack of systolic left ventricular contours. (b) Polygon recontracted from systolic contours in (a) compared with diastolic contours from the same level. Reproduced from *Radiology*[6], with permission.

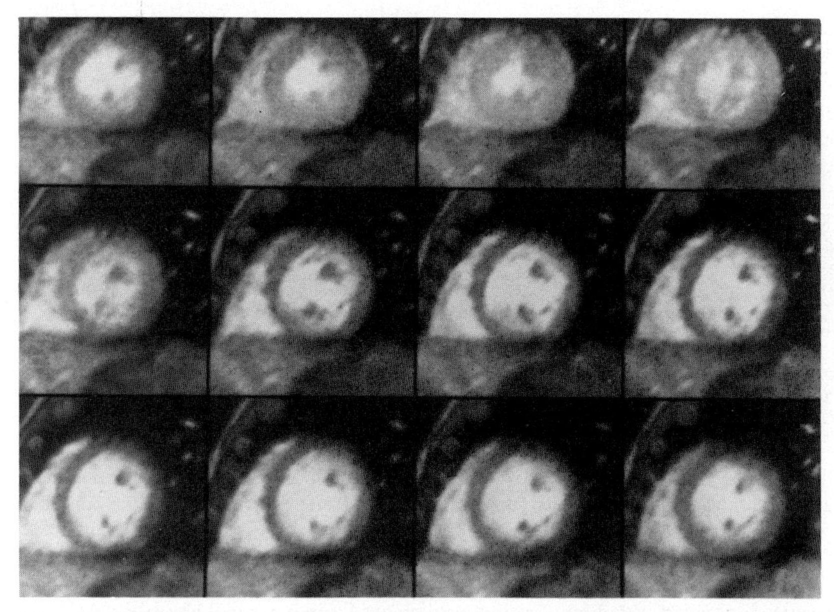

Figure 5 Cine MR images of a short axis view of the heart recorded with gradient-refocused echoes. Note the high intensity of blood in the chambers, differing from Fig. 3 recorded with spin echo techniques. Note the thicker myocardial wall in systole (Frame 3) as opposed to diastole (Frame 10). (Photograph courtesy of Dr Richard D. White, Cleveland Clinic Foundation).

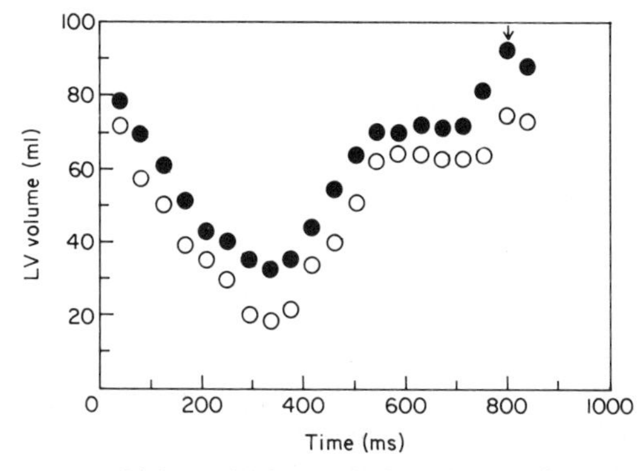

Figure 6 Right and left ventricular volumes of a normal subject recorded over the R–R interval. Stroke volumes may be obtained by subtraction of the volume at systole from the volume at diastole. Arrow marks filling due to atrial contraction. ○ = LV volume (ml); ● = RV volume (ml). Adapted from ref. 8.

Ventricular cavities can be outlined and the total volume calculated by adding the outlined areas from images from the entire heart and multiplying by the section thickness. The smallest volume represented the systolic volume and the largest indicated diastolic volume, the difference representing stroke volume. The entire time series of the ventricular volumes of a normal subject[8] is shown in Fig. 6.

The previous studies have all been performed at rest since MRI is particularly sensitive to patient motion, which complicates measurements made during exercise. Recent studies, however, have been performed with dipyridamole as a pharmacologic stressor. In this report[9], short axis views of the heart are shown in diastole and systole before and after dipyridamole administration. Although coronary arteriography showed normal global and regional left ventricular function, the anteroseptal hypokinesis and reduced thickening of the myocardium seen at systole in this segment was consistent with the blocked proximal left anterior descending artery in this patient.

These stress techniques may extend the usefulness of MRI not only to new measurements of wall motion and myocardial thickness but also to volume changes induced with stress.

Cine-computed tomography (cine-CT)

Computed tomography was initiated by rotating both the detectors and X-ray source 360° around the body[1]. Later techniques had speeded this process considerably by using a stationary ring of detector with a mechanically rotating X-ray tube.

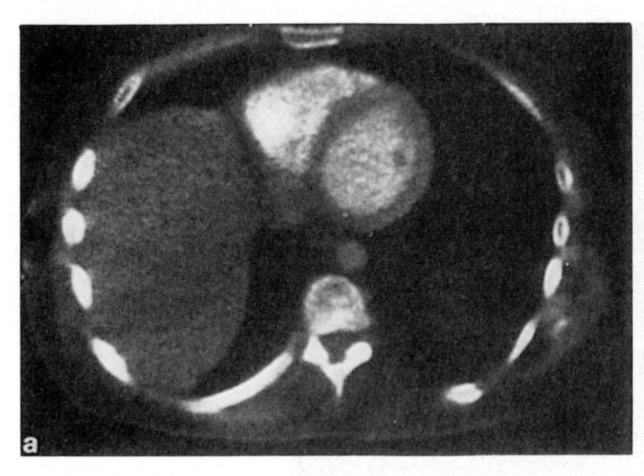

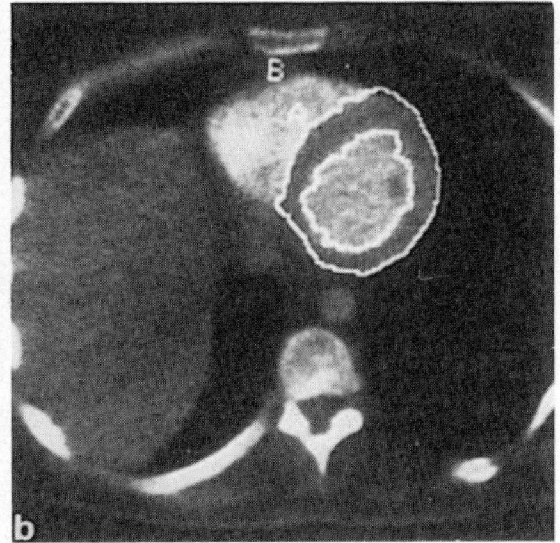

Figure 7 (a) A 50 ms cine-CT image taken in the short axis from a series of parallel scans obtained at the apex (b) outline of LV myocardium as recorded at each cardiac level and by which summation will yield left ventricular mass. Reproduced from *Radiology*[12], with permission.

Cine-CT has provided greatly increased speed by varying the position of the X-ray beam by magnetic deflection, so that an 8 mm tomographic section can be recorded in 50 ms[10]. Essentially eight sections (7·6 cm including interslice spacing) can be imaged by scanning four tungsten targets with no mechanical motion involved[11].

The geometric applications are quite similar to the MRI techniques except that a higher resolution can be obtained by use of matrices for 256×256 to 512×512. As shown in Fig. 7, the use of contrast material allows a well-defined differentiation between the chamber and myocardium so that the volume within a series of contours can be summated to yield total chamber volume or total chamber mass[12].

Again, subtraction of the volumes calculated at diastole and systole will yield stroke volume and cardiac output by combining with the heart rate. Aortic regurgitation can be derived also by determining the difference between right and left stroke volumes[13]. Although this technique involves multiple subtractions of systolic volumes from diastolic volumes in both right and left ventricles, correlation of ventricular volumes with electromagnetic flowmeters or thermodilution techniques have been estimated with standard errors as low as 1·4–1·7 ml. Further subtraction showed the differences between right and left stroke volumes to average 1·1 ml with a range of 0·1–3·2 ml[14].

Dilution curves may be obtained with the cine-CT method by measuring the varying intensity (Hounsfield units) of the contrast material as it enters and clears from a central chamber. Multiple frames can be recorded within a period of 1 s and can be analysed by the Stewart-Hamilton equation to calculate cardiac output[15] from curves shown in Fig. 8[16]. Time concentration curves have been derived from cine-CT to determine patency of coronary artery bypass grafts[17] and have also been recorded before and after dipyridamole to show flow rates and flow reserve[18].

Discussion

The purpose of this report has not been to compare the specifications of these three imaging modalities or to review the many applications that have been performed with them. It has been instead to discuss the characteristics that may make these imaging systems applicable to measurements of cardiac volumes either now or in the future.

It must be kept in mind that these systems have all been designed for imaging, but can be used, as can the scintillation camera which was also designed for imaging, as a highly collimated probe to obtain a time-concentration curve of an administered tracer material. In terms of basic imaging considerations, there is no question that, assuming a differentiation can be made between the endocardium and the blood in the cardiac chambers, a series of contours can be summated to yield volumes, stroke volumes, cardiac ouput and regurgitation volumes. Other characteristics such as wall motion, myocardial thickness, and changes in myocardial thickness as a function of contraction, filling, or stress can all be

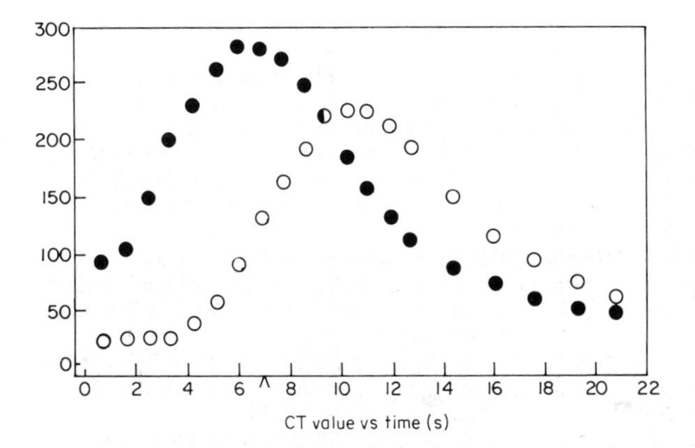

Figure 8 Dilution curves from regions of interest placed over the left atrium (●) and descending aorta (○) recorded at 1 s intervals following a 35 ml bolus of 76% iodinated contrast material injected intravenously at a rate of 5 ml per second. Adapted from ref. 16.

measured. At present only MRI and cine-CT have demonstrated the spatial resolution necessary for clinical utility in this respect.

All three of the modalities can yield time-concentration curves. At present, both PET and cine-CT involve tracers which have demonstrated linearity in the measurement of concentration. In the conventional flow applications of MRI, a tracer is not used but differentiation of moving blood from myocardium has been achieved by the choice of imaging techniques. It has been possible to correlate signal intensity with flow rate in isolated systems but it is difficult to extrapolate this relationship to physiological measurements. It has been possible to calibrate MRI flow rate by use of gadolinium contrast material[19] to measure the response of flow to dipyridamole; however, prolonged administration of the contrast material was used rather than a short bolus.

What is perhaps a more significant consideration is whether the information from the new imaging modalities is superior to other methods, e.g. are the cardiac output values derived from stroke volumes calculated from summation of slices more accurate than those from dilution curves? A second consideration is the cost of the measurement and whether the advantages of an atraumatic technique justify its expense. In most of the evaluations cited, the results with the new method were judged in comparison with those of existing techniques, such as thermodilution. The instrument requirements may be vastly different but the value can only be assessed for specific applications, i.e. the need for repeated measurements, non-invasive characteristics, and availability of results for interventional procedures.

At present, it is pertinent only to define what procedures can be done, while the specific application will ultimately determine which technique will be used.

Finally, there are many areas currently being investigated with these modalities that have not been discussed here since they are not directly related to cardiac volumes. Magnetic resonance angiography, new PET radiopharmaceuticals, and the several proposed measurements of myocardial flow reserve however could all be useful in cardiac evaluation. It will be important to continue to survey the advances made with these imaging systems for possible incorporation into future studies.

Conclusions

With respect to cardiac volumes by the geometric method, the present resolution of PET indicates this modality has little to offer at present. Use of in vivo-derived arterial dilution curves for absolute quantification of myocardial perfusion with PET may be very important, however, and could be used for dilution curve cardiac ouput measurement if desired.

MRI and cine-CT provide an elegant geometric method for measurement of ventricular volumes, stroke volumes, and aortic regurgitation and cardiac output derived from the reconstructed volumes. The accuracy appears similar to alternative methods and may justify the added expense or some studies, but a need for such accuracy has not yet been defined. The same method can be applied to measurement of ventricular mass which may be more significant.

The large volume required for bolus injections with cine-CT and possibly MRI would make cardiac output measurements using the dilution method with these modalities more difficult than the existing methods. The application of dilution curves to measure myocardial perfusion appears promising on all three of the modalities.

All three of these modalities are in the process of rapid growth and should be continuously monitored to determine applicability of these new techniques to various cardiac measurements.

References

[1] Hounsfield GN. Computerized transverse axial scanning (tomography) Part I. Description of system. Br J Radiol 1973; 46: 1016–22.

[2] Go RT, Marwick TH, MacIntyre WJ *et al.* A prospective comparison of rubidium-82 PET and thallium-201 SPECT myocardial perfusion imaging utilizing a single dipyridamole stress in the diagnosis of coronary artery disease. J Nucl Med 1990; 31.

[3] Marwick TH, MacIntyre WJ, Go RT, Underwood D, Saha GB. Assessment of ischemic burden with post exercise fluorodeoxyglucose positron emission tomography. Eur J Nucl Med 1990; 16: 472.

[4] Weinberg IN, Huang SC, Hoffman EJ *et al.* Validation of PET-acquired input functions for cardiac studies. J Nucl Med 1988; 29: 241–7.

[5] MacIntyre WJ, Go RT, Feiglin DH, O'Donnell JK. Cardiac magnetic resonance imaging: Basic principles and clinical potential. In: Freeman LM and Weissman HS, Eds. Nuclear medicine annual 1985, New York: Raven Press 1985: 171–98.

[6] Edelman RR, Thompson R, Kantor H, Brady TJ, Leavitt M, Dinsmore R. Cardiac function: Evaluation with fast-echo imaging. Radiology 1987; 162: 611–5.

[7] Sechtem U, Pflugfelder PW, Gould RG, Cassidy MM, Higgins CB. Measurement of right and left ventricular volumes in healthy individuals, with cine MR imaging. Radiology 1987; 163: 697–702.

[8] Sechtem U, Pflugfelder PW, White RD *et al.* Cine MR imaging: Potential for the evaluation of cardiovascular function. Am J Radiol 1987; 148: 239–46.

[9] Pennell DJ, Underwood SR, Longmore DB. Detection of coronary artery disease using MR imaging with dipyridamole infusion. J Comput Assist Tomogr 1988; 14: 167–70.

[10] Boyd DP, Gould RG, Quinn JR, Herrmansfeldt W. A proposed dynamic cardiac 3-D densitometer for early detection and evaluation of heart disease. IEEE Trans Nucl Sci 1979; 26: 2724–7.

[11] Lipton MJ, Higgins CB, Farmer D, Boyd DP. Cardiac imaging with a high speed cine-CT scanner: Preliminary results. Radiology 1984; 152: 579–82.

[12] Diethelm L, Simonson JS, Dery R, Gould RG, Schiller NB, Lipton MJ. Determination of left ventricular mass with ultrafast CT and two-dimensional echocardiography. Radiology 1989; 171: 213–7.

[13] Reiter SJ, Rumberger JA, Standford W, Marcus ML. Quantitative determination of aortic regurgitant volumes in dogs by ultrafast computed tomography. Circulation 1987; 76: 728–35.

[14] Reiter SJ, Rumberger JA, Feiring AJ *et al.* Precision of right and left ventricular stroke volume measurements by rapid acquisition cine computed tomography. Circulation 1986; 74: 890–900.

[15] Garrett JS, Lanzer P, Jaschke W *et al.* Measurement of cardiac output by cine computed tomography. Am J Cardiol 1985; 56: 657–61.

[16] Steiner RM, Flicker S, Eldredge WJ *et al.* Clinical experience with rapid acquisition cardiovascular CT imaging (cine CT) in the adult patient. Radiographics 1989; 9: 283–305.

[17] Stanford W, Brundage BH, MacMillan R *et al.* Sensitivity and specificity of assessing coronary bypass graft patency with ultrafast computed tomography: Results of a multicenter study. J Am Coll Cardiol 1988; 12: 1–7.

[18] Rumberger JA, Feiring AJ, Hiratzka LF *et al.* Quantification of coronary artery bypass flow reserve in dogs using cine computed tomography. Circ Res 1987; 61 (Suppl 2): 117–23.

[19] Miller DD, Holmvang G, Gill JB *et al.* MRI detection of myocardial performance changes by gadolinium DTPA infusion during dipyridamole hyperemia. Magn Reson Med 1989; 10: 246–55.

Magnetic resonance imaging and cine computerized tomography as future tools for cardiac output measurement

P. OMVIK

Medical Department, Section of Cardiology, Haukeland Hospital, Bergen, Norway

KEY WORDS: Cine computerized tomography, magnetic resonance imaging.

The new imaging methods, cine computerized tomography (cine-CT) and magnetic resonance imaging (MRI), are reviewed as tools for cardiac output (CO) measurement. Both modalities provide accurate and highly reproducible measurements of stroke volume and CO at rest. The methods are less applicable during heavy physical exercise. Because of cost and lack of equipment the methods are not practical for measurements of CO alone in the foreseeable future. However, cine-CT and MRI may be of value for CO measurements when additional characteristics, such as wall motion, myocardial structure and tissue perfusion, need to be assessed.

Introduction

In normal adult man, cardiac output (CO) varies from approximately 5 l min^{-1} at rest to 30 l min^{-1} during maximal physical work. In cardiac disease, CO usually is reduced, particularly during exercise. Thus, an optimal method for CO determination should provide accurate measurements over a wide range of CO at rest as well as during exercise. The measurement should be fast and convenient, and it should be repeatable at short intervals. The reproducibility of the measurement should be high and the variability low. Finally, its cost/benefit ratio should be low.

This is a critical review of two new imaging techniques, cine computerized tomography (cine-CT) and magnetic resonance imaging (MRI), as tools for CO measurement. Both methods provide rapid sequential or gated images of the left (and/or right) ventricle from which stroke volume (SV) and CO are calculated[1–4]. Complex geometric shapes of the ventricle can be allowed for by using multiple sections. By computer techniques, three-dimensional display of the ventricular cavity can be constructed[5]. The CO may also be calculated from dilution curves obtained by cine-CT[4,6].

Cine computerized tomography

The use of computerized tomography for CO measurements was made possible by the construction of a new, ultrafast cine-CT which provides simultaneous, multilevel millisecond X-ray scanning of the heart[1,3,6]. The scanning can include the entire ventricle for one cardiac cycle. Thus, no ECG gating is required. The CO may be calculated by time-density curves, using the Stewart-Hamilton equation, or from geometrically determined SV[1,3,5,6]. Cine-CT is not completely non-invasive since contrast agents have to be injected.

Magnetic resonance imaging

Magnetic resonance imaging of the ventricular cavity is based on different proton radiofrequency emission density in myocardium and flowing blood[2–4]. The method is completely non-invasive and there is no use of ionized radiation. Thus, except for the need for adequate acquisition time (30 min) there is virtually no limit on repetition of measurements.

Advantages of cine-CT and MRI for CO measurements

The accuracy of CO measurements by cine-CT and MRI has been reported to be high, with correlation coefficients >0·90 when compared with other (invasive) methods[6–9]. Still, large differences (e.g. 2 l min^{-1}) between measurements by MRI and the reference method have been seen[6]. However, correlation analysis is not the best way of comparing two methods of clinical measurement; a statistically superior technique is assessment of the limits

Address for correspondence: Per Omvik, M.D., Section of Cardiology, Medical Department, Haukeland Hospital, 5021 Bergen, Norway.

of agreement between the two measurements and, for a given CO, the mean difference between CO measured by cine-CT or MRI and the chosen reference method. In a comparison between MRI and indicator dilution, Møgelvang *et al.* found a mean difference of −6 ml (with limits of agreement −22–10 ml, or about −25–12%) between SV determined by the two methods[8].

Reproducibility of CO measurements by cine-CT has been reported to be extremely good[3,4,7], and the reproducibility of measurements by MRI is also high[3,4]. For both methods consecutive measurements generally differ by $<5\%$ and the correlation coefficient between paired measurements is $>0{\cdot}95$. Similar reproducibility is usually shown by invasive measurements using cardiogreen, a standard of CO measurement[10], but is not commonly achieved with other non-invasive methods. By both methods, intra- and inter-individual variability is low, with a coefficient of variation $<10\%$[4,7].

In addition to measurements of CO, the two new imaging modalities, cine-CT and MRI, provide detailed information on other characteristics of cardiac function, e.g. wall motion, wall thickness and myocardial perfusion[1,4] These are fields for future application of CO measurement by the two methods. For instance, in hypertensive patients the methods could be used to relate systolic and diastolic left ventricular wall motion to simultaneously measured SV and CO. In ischaemic heart disease it could be of interest to relate myocardial perfusion to CO. Also, by a third imaging modality, positron emission tomography (PET), CO theoretically might be derived from dilution curves[4]. Combined with positron emission from metabolic tracers, use of PET might provide new insights into biochemical changes related to the cardiac cycle and CO in patients with ischaemic heart disease.

Some of these features could be investigated by more common non-invasive methods such as echocardiography[3]. Nevertheless, an adequate echocardiographic image cannot be obtained from every patient and, until now, measurements have been based on M-mode recordings or single plane two-dimensional display of the ventricle. Moreover, by echocardiography, the border between endocardium and the ventricular cavity is often less precisely defined than on cine-CT or MRI[1,2,5].

All of these new imaging modalities, including cine-CT and MRI, PET and echocardiography, provide information on myocardial structure, geometry and function. There is considerable overlap in performance between the methods; however, the relative place of each in analysis of cardiac function remains to be established.

Table 1 Advantages and disadvantages of magnetic resonance imaging for cardiac output measurement

The main advantages of MRI:
No ionizing radiation
No contrast media
Imaging performed in any plane
Additional cardiac characterization
Versatile technique
The main disadvantage of MRI:
High cost
Long imaging time
Claustrophobic effect
Difficulty in sick patients
Limited use during exercise

Drawbacks of cine-CT and MRI for CO measurement

The single most important disadvantage of MRI for CO measurement is the cost, which by far surpasses that of other methods[3]. Because of lack of MRI equipment, measurements of CO by MRI today can be performed only in a few, selected hospitals. At present fewer than 50 ultrafast cine-CT machines have been installed. Although the cost of cine-CT equipment is less than that of MRI, the cost/benefit ratio of cine-CT for CO measurement is still far greater than that of conventional methods[1,3].

Today, technical progress in cardiac imaging is rapid[3,4], and it is likely that the cost of equipment both for MRI and cine-CT will be greatly reduced. However, even when equipment for the new imaging modalities becomes more generally available, running cost will remain extremely high for the foreseeable future.

Another limiting factor for CO measurements by cine-CT and MRI is the size of the equipment[1,2]. As neither cine-CT nor MRI equipment can be moved, the patient has to be studied at the site of the machines, in contrast to other non-invasive (e.g. echocardiography) and invasive methods which may be applied at the bedside and in emergency situations[3].

Cardiac output is a dynamic variable that is strongly influenced by the autonomic nervous system. By using cine-CT or MRI, the patient is literally placed *within* the machine. In this situation, claustrophobia might well alter the measurements.

The construction of the MRI machine and the need for sequential emission sampling also preclude CO measurements by MRI during vigorous dynamic exercise, which is essential for studies of the functional capacity of the heart. A summary of advantages and drawbacks of MRI for CO measurements is shown in Table 1. With cine-CT attempts have been made to obtain scans of the left ventricle during dynamic exercise[11]. However, the patient still has to maintain a nearly fixed chest position during exercise, and must be supine or slightly (20°) axially tilted[11].

Conclusion

Accurate and reproducible measurements of CO at rest may be obtained by cine-CT and MRI, but because of cost and lack of equipment measurement of CO alone will not be practical on a general basis for the foreseeable future. Measurement during exercise is limited by chest movement and semi-supine position. In selected patients, the methods may be of value for CO measurements when such additional characteristics of cardiac function as wall motion and myocardial structure are to be determined. The relative roles of the different imaging modalities for evaluation of cardiac function, including CO, have yet to be determined.

References

[1] Boyd DP, Couch JL, Napel SA, Peschmann KR, Rand RE. Ultra cine-CT for cardiac imaging: Where have we been? What lies ahead? Am J Cardiac Imaging 1987; 1: 175–85.

[2] Pohost GM, Canby RC. Nuclear magnetic resonance imaging: current applications and future prospects. Circulation 1987; 75: 88–95.

[3] Roelandt J, Sutherland GR, Hugenholtz PG. The 1980s renaissance in cardiac imaging: The role of ultrasound. Eur Heart J 1989; 10: 680–4.

[4] MacIntyre WJ. Potential application of new imaging modalities in the measurement of cardiac volumes and characteristics. Eur Heart J 1990; 11 (Suppl I): 133–40.

[5] Skorton DJ, Collins SM. Evolving cardiac applications of digital processing in ultrasound, ultrafast CT, and magnetic resonance. Am J Cardiac Imaging 1988; 2: 3–15.

[6] Garrett JS, Lanzer P, Jaschke W *et al.* Measurement of cardiac output by cine computed tomography. Am J Cardiol 1985; 56: 657–61.

[7] Reiter SJ, Rumberger JA, Feiring AJ, Stanford W, Marcus ML. Precision of measurements of right and left ventricular volume by cine computed tomography. Circulation 1986; 74: 890–900.

[8] Culham JAG, Vince DJ. Cardiac output by MR imaging: an experimental study comparing right ventricle and left ventricle with thermodilution. J Can Assoc Radiol 1988; 39: 247–9.

[9] Møgelvang J, Stubgaard M, Thomsen C, Henricksen O. Evaluation of right ventricular volumes measured by magnetic resonance imaging. Eur Heart J 1982; 9: 529–33.

[10] Lund-Johansen P. The dye dilution method for measurement of cardiac output. Eur Heart J 1990; 11 (Suppl I): 6–12.

[11] Roig E, Chomka EV, Castaner A *et al.* Exercise ultrafast computed tomography for the detection of coronary artery disease. J Am Coll Cardiol 1989; 13: 1073–81.

Validation of non-invasive measurement of cardiac output. The Ann Arbor experience

S. Julius

Division of Hypertension, Department of Internal Medicine, University of Michigan Medical School, Ann Arbor, Michigan, U.S.A.

KEY WORDS: Echo-Doppler, CO_2 rebreathing, M-mode echocardiography.

In an assessment of non-invasive methods for the clinical measurement of cardiac output, both M-mode echocardiography and the CO_2 rebreathing technique were found to be unreliable. However, the echo-Doppler method appeared for population studies.

Introduction

The utilization of the dye dilution and thermodilution, the two standard methods of haemodynamic measurement, is severely limited by their invasive nature. Only relatively small groups of patients can be investigated, studies are performed in very specialized laboratories and the material seen in such laboratories is usually highly preselected. These limitations are of particular concern in haemodynamic studies on patients with borderline hypertension. Over a period of years we[1,2] and others[3,4] have consistently found a state of hyperkinetic circulation among young subjects with borderline hypertension. These subjects have an increased sympathetic stimulation and a decreased parasympathetic inhibition of the heart[2]. The question arises of whether this aberration in the autonomic tone reflects a temporary emotional response to conditions of measurement or whether it is characteristic of these individuals under normal circumstances. Furthermore, it is not known whether self-referred patients seen in the hospitals are representative of subjects with borderline hypertension at large.

These issues could be resolved only through the development of a non-invasive haemodynamic method which could be utilized on large populations outside of specialized laboratories. Consequently we undertook to evaluate various methods of non-invasive haemodynamic measurements as they became available. It will be shown that simple M-mode echocardiography and CO_2 rebreathing methods did not yield sufficiently reliable or sensitive data. However, the results with 2D echo combined with Doppler sonography of the aortic outflow were encouraging and led to the implementation of that method in a large epidemiologic study.

Address for correspondence: Stevo Julius, M.D., Sc.D., University of Michigan, Division of Hypertension, 3918 Taubman Center, Ann Arbor, MI 48109–0356, U.S.A.

Supported by grant HL37464 from the National Heart, Lung and Blood Institute, NIH, Bethesda, MD, USA.

Results

The CO_2 rebreathing was examined in an early study[5] and revisited in a more recent investigation[6]. In the first study we used a modification of the Defares method[7] and compared it with dye dilution in 14 healthy normal men aged 26 ± 6·6 years. The CO_2 rebreathing produced reasonable within-test reproducibility. Correlation coefficients of two repeated determinations of cardiac output were 0·82 at rest, 0·93 at moderate exercise and 0·86 at a high exercise level. Compared with dye dilution, CO_2 rebreathing yielded fair results only during exercise. At rest, the correlation with dye dilution was not significant ($r = 0{\cdot}22$), but with exercise the congruence of the two methods improved; the correlation coefficient was 0·54 during mild, 0·64 at moderate and 0·87 at maximum exercise. Clearly, the larger the exercise-induced A/V CO_2 difference, the better the result. Consequently, at that time, CO_2 rebreathing did not appear a sufficiently reliable method for haemodynamic measurement. Later, the CO_2 rebreathing method with Collier's equilibration technique to measure mixed pCO_2 yielded somewhat better, but still not fully satisfactory, results[8].

Our next step was to validate M-mode echocardiography as a tool for non-invasive measurements of the stroke volume[9]. The study was performed on 10 normotensive volunteers at two periods of rest,

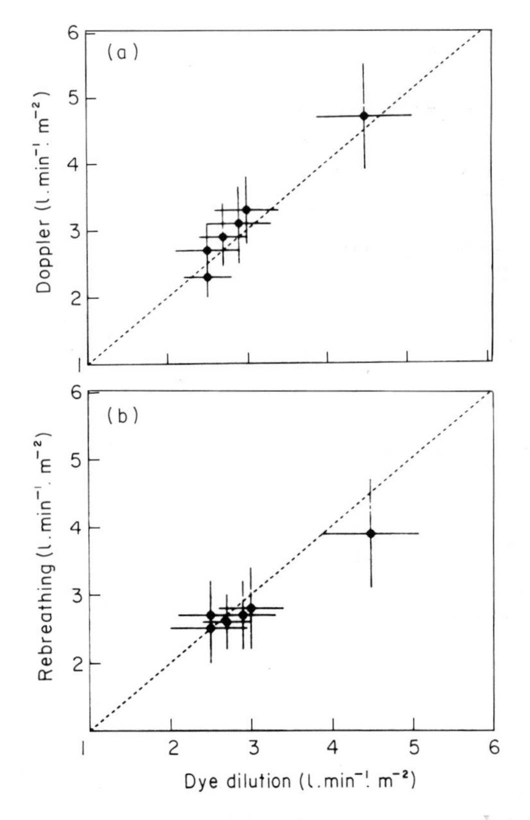

Fig. 1 Relationship between the mean cardiac output values as measured by dye dilution and (a) echo-Doppler and (b) CO_2 rebreathing for six conditions of measurements. Cross-hatched lines represent the standard error of the mean for each method. Reproduced by permission from Clin Pharmacol Ther 1987; 41: 419–25.

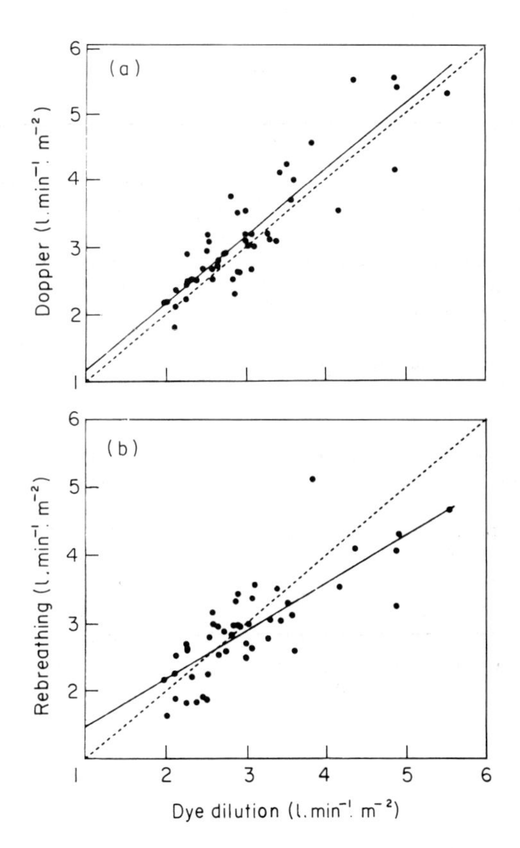

Fig. 2 Relationship between each cardiac output value measured by dye dilution and (a) echo-Doppler and (b) CO_2 rebreathing. The dashed line is the line of identity, whereas the solid line represents the least square regression line. Reproduced by permission from Clin Pharmacol Ther 1987; 41: 419–25.

after infusion of 15 mg kg^{-1} min^{-1} of isoproterenol and after injection of 0·2 mg kg^{-1} of propranolol. The overall trend in cardiac output determined by dye dilution and by M-mode echocardiography was similar by both methods. The first resting cardiac output was 6·21 ± SE 0·32 l min^{-1} by dye dilution vs 5·67 by echocardiography. The second resting cardiac output was 6·29 ± 0·48 vs 6·22 ± 0·61 by echocardiography. Similarly, the values after propranolol were comparable; 4·87 ± 0·33 by dye dilution vs 4·65 ± 0·37 by echocardiography. However, the cardiac output after isoproterenol was larger when assessed by dye dilution (10·53 ± 0·67 l min^{-1}) than by echocardiography (8·65 ± 0·72 l min^{-1}). A detailed analysis of results with isoproterenol showed that echocardiography can depict an increase of stroke volume up to 40% over baseline, but above 40% the estimates of stroke volume by echocardiography are substantially lower than by dye dilution. Furthermore, while the average resting values of cardiac output were similar with both methods, the correlation of the two methods was poor; r = 0·52, P = 0·1 first rest period and r = 0·72, P = 0·02 second rest period. Finally, correlation between two consecutive measurements of cardiac output with the same method favoured dye dilution (r = 0·94) over echocardiography (r = 0·85). We considered these results discouraging and decided that M-mode echocardiography could not be used as a tool for haemodynamic research.

Encouraged by reports in the literature, we undertook to evaluate the practical utility of the echo-Doppler method of assessment of cardiac output[6]. The more promising CO_2 rebreathing method with the Collier technique was also evaluated. The study was performed on eight male healthy

Table 1 Mean cardiac index changes in response to cuff inflation, tilt, isoproterenol, and propranolol as measured by each method, with calculated sample sizes needed to detect changes at the 5% level

Intervention	Dye dilution ($l.min^{-1}.m^{-2}$)	Doppler ($l.min^{-1}.m^{-2}$)	CO_2 rebreathing ($l.min^{-1}.m^{-2}$)
Cuff			
Mean cardiac index change	−0·36*	−0·43*	−0·18
SD	0·22	0·21	0·42
Sample size	<5	<5	33–40
Tilt			
Mean cardiac index change	−0·59*	−0·95	−0·06
SD	0·37	0·36	0·38
Sample size	<5	<5	>101
Isoproterenol			
Mean cardiac index change	−1·59*	−1·67*	1·21*
SD	0·64	0·49	0·71
Sample size	<5	<5	<5
Propranolol			
Mean cardiac index change	−0·45*	−0·39*	−0·22
SD	0·29	0·22	0·42
Sample size	5	<5	22–27

Estimated sample sizes are based on each technique's mean change and SD, with α set at 0·05 and β set of 0·2. *$P<0·005$.

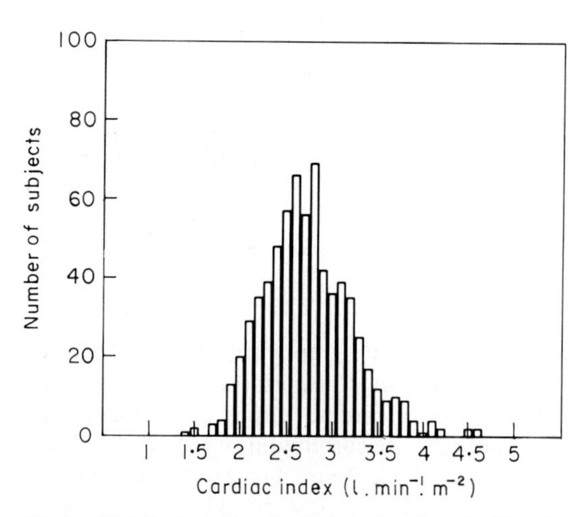

Fig. 3 Distribution of cardiac index values by echo-Doppler methodology in 692 subjects in Tecumseh, Michigan.

volunteers under the following conditions: (1) supine rest; (2) after inflation of cuffs around the thighs; (3) 35° head-up tilt; (4) repeated supine rest; (5) during constant isoproterenol infusion and (6) after an intravenous bolus of propranolol. Three manipulations known to decrease cardiac output were chosen for their disparate haemodynamic effects: tilt lowers output and increases heart rate; propranolol decreases both the cardiac output and heart rate; whereas thigh cuff inflation decreases cardiac output and has very little effect on heart rate. Isoproterenol infusion added a point of high cardiac output measurement. The results for the whole group are given in Fig. 1. Both echo-Doppler and CO_2 rebreathing gave similar group values. Individual data points are shown in Fig. 2, and it can be seen that the CO_2 rebreathing results correlate with dye dilution results less strongly than do those of echo-Doppler with those of dye dilution. Furthermore, CO_2 measurement of cardiac output tended to underestimate cardiac output at higher levels. Generally the CO_2 method showed a larger variance and was less sensitive in detecting differences. To evaluate the clinical applicability of these two methods, we calculated the sample size estimates needed to detect differences seen in our study if one were to use each of these methods. These data are shown in Table 1. It can be seen that the CO_2 rebreathing technique is not reliable, whereas the changes of cardiac output could be detected on an equally small number of subjects by the echo-Doppler and the dye dilution methods. We concluded that, in every regard, measurements of cardiac output by echo-Doppler are reliable and the method can be utilized for studying a larger number of subjects.

We recently performed a large number of cardiac output measurements in the village of Tecumseh, Michigan. As shown in Fig. 3, the values were distributed in a classic bell-shaped manner. We retrieved from the records resting measurements on normotensive and hypertensive subjects investigated over several years in our laboratory[10]. The correlation coefficient of cardiac output with body size in these 280 subjects is compared with the correlation obtained by the echo-Doppler technique in 774 subjects in Tecumseh. Weight correlated with cardiac output by dye dilution with $r = 0{\cdot}39$, whereas, for echo-Doppler, $r = 0{\cdot}45$. Correlation of the cardiac output with height for dye dilution was $r = 0{\cdot}29$ and for echo-Doppler, 0·27.

Conclusions

There is a great need to develop good non-invasive methods for measurement of cardiac output. Over the years we have searched for such a method but our results with M-mode echocardiography and with CO_2 rebreathing were disappointing. A detailed study under controlled circumstances and performed with special care suggested that the echo-Doppler method may be reliable and sensitive. This led to a field testing of the method in the context of a large epidemiologic study. The cardiac index values obtained were reasonable and showed a normal distribution. The cardiac output correlated to body size in a similar manner by echo-Doppler to that found by dye dilution measurements. We believe that the echo-Doppler method is adequate for large-scale population studies of haemodynamics.

References

[1] Julius S, Conway J. Hemodynamic studies in patients with borderline blood pressure elevation. Circulation 1968; 38: 282–8.

[2] Julius S, Pascual AV, London R. Role of parasympathetic inhibition in the hyperkinetic type of borderline hypertension. Circulation 1971; 44: 413–8.

[3] Lund-Johansen P. Hemodynamics in early essential hypertension. Acta Med Scand 1967; 482 (Suppl): 1–105.

[4] Sannerstedt R. Hemodynamic response to exercise in patients with arterial hypertension. Acta Med Scand 1966; 458 (Suppl): 1–83.

[5] Ferguson RJ, Faulkner JA, Julius S, Conway J. Comparison of cardiac output determined by CO_2 rebreathing and dye-dilution methods. J Appl Physiol 1968; 25: 450–4.

[6] Hinderliter AL, Fitzpatrick MA, Schork N, Julius S. Research utility of noninvasive methods for measurement of cardiac output. Clin Pharmacol Ther 1987; 41: 419–25.

[7] Defares JG. Determination of $PvCO_2$ from the exponential CO_2 rise during rebreathing. J Appl Physiol 1958; 13: 159–64.

[8] Collier CR, Affeldt JE, Farr AF. Continuous rapid infrared CO_2 analysis. J Lab Clin Med 1955; 45: 526–39.

[9] Kiowski W, Randall OS, Steffens TG, Julius S. Reliability of echocardiography in assessing cardiac output. A comparative study with a dye dilution technique. Klin Wochenschr 1981; 59: 1115–20.

[10] Julius S, Schork N, Schork A. Sympathetic hyperactivity in early stages of hypertension: The Ann Arbor data set. J Cardiovasc Pharmacol 1988; 12 (Suppl): S121–9.

Clinical assessment of cardiac output

J. CONWAY

Cardiac Department, John Radcliffe Hospital, Headington, Oxford, U.K.

KEY WORDS: Techniques, accuracy, repeatability.

Cardiac output estimation is an important and much needed measurement for assessing patients in heart failure. In hypertension, it is vital for understanding the haemodynamic basis of the disease and the mode of action of drugs. Measurements of blood pressure and cardiac output provide the only means of estimating peripheral resistance. Of the available methods to determine cardiac output, thermodilution is the most practical, although it has its difficulties and care has to be exercised in its use. When intra-arterial blood pressure measurements are needed, the dye-dilution method is equally valid, and if respiratory techniques are available the Fick principle may also be used.

Of the non-invasive methods, none is yet developed to a stage suitable for general clinical use. Doppler velocimetry is the most promising technique, but it requires complex computer analysis and, as yet, can reliably be used only to measure changes in cardiac output in an individual. The technique has been assessed against the electromagnetic flowmeter in man and gives reasonable accuracy and repeatability. Echocardiography and impedance cardiography are not yet satisfactory for clinical use; neither are the radionuclide methods, apart from the 'first pass' method, but this also needs further verification.

Introduction

The appropriate method for the measurement of cardiac output depends very much on the intended use of the resulting information. Clinicians have priorities that differ from those of clinical investigators: the latter are trained to be critical, to assess evidence and to examine the limitations of their techniques; clinicians, on the other hand, in general want to use a method for measuring cardiac output as a tool in decision making, either for diagnosis or for treatment. Clinicians nevertheless often need to publish data, so their methods must be acceptable. The choice of method depends to some extent upon the expertise available in a particular department and upon the clinical circumstances in which measurements are to be made.

In clinical cardiology, cardiac output estimations are not often needed in diagnosis, partly because background information is not available owing to the lack of easy methods for clinical use. However, cardiac output measurements are often essential in assessing progress and, particularly in heart failure, in monitoring the response to treatment. In myocardial infarction and in shock the technique is needed to assess progress.

In hypertension there are three main requirements for the measurement of cardiac output. The first of these is to verify the existence of a 'hyperdynamic state'. In spite of the large body of evidence in favour of raised cardiac output in some forms of hypertension, uncertainty persists because the presence of a high cardiac output may merely be evidence for an exaggerated alerting response to the laboratory situation.

Secondly, information on cardiac output is needed to understand the mechanisms operating in hypertension and to enable one to estimate the level of peripheral resistance. For example, the notion of a 'high resistance' type of hypertension cannot be considered seriously until it is based on accurate measurements of cardiac output as well as of blood pressure.

Thirdly, in investigating the mechanism of hypotensive agents, knowledge of cardiac output is essential. For example, analysis of the mode of action of diuretics initially suggested that there was a fall in cardiac output, which seemed the logical effect of a depletion in blood volume. However, it was later shown that this phase lasts but a few weeks and it is followed by a fall in peripheral resistance, which persists.

Address for correspondence: James Conway, Cardiac Department, John Radcliffe Hospital, Headington, Oxford OX3 9DU, U.K.

Which method to recommend?

As a general rule, the easier the method the more appealing it is to those who need a tool, but the more likely it is to be abused by uncritical observers. The literature is replete with results of untested methods; examples can be quoted from ballistocardiography, ear oximetry etc. Here, we have discussed problems with more contemporary methods: impedance cardiography has been extensively used and yet we learn that the empirical equations upon which it is based require revision; Doppler has been used and yet arguments still persist on the precise methodology to be used; the same applies to echocardiography. There is no simple 'push button' method for accurately measuring cardiac output. We have learnt from the papers on indicator dilution methods and the Doppler technique that the operator must be presented with the primary data for visual examination in order to verify the quality of the data. There are, in fact, no methods available which can be used by untrained personnel. That being said, there are methods which are eminently suitable for measurement of cardiac output.

The Fick principle is valid and accurate and would be recommended to clinicians who had the appropriate respiratory techniques readily available. The indirect Fick method based on carbon dioxide appears to be reasonably accurate during exercise when large amounts of carbon dioxide are being produced by muscle, but it is not acceptable at rest.

Dye-dilution has been the basis of accurate determination of cardiac output for many years. It has a repeatability of 5% and has replaced the Fick method as a 'standard'. It requires venous catheterization (with injection in or near the right atrium) and arterial catheterization with a catheter of sufficient size to allow for the withdrawal of blood at 15–20 ml min^{-1}. It is the ideal method when accurate arterial pressure measurements are also needed, thereby justifying arterial puncture. This applies particularly during exercise when indirect measurements of blood pressure are notoriously inaccurate.

Thermodilution is the most convenient indicator dilution method for clinical use, but it does require rigid adherence to certain requirements. Right-sided catheterization to the pulmonary artery with injection into the right atrium is necessary. The positioning of a thermistor in the right atrium is recommended and powered injection of 10 ml of ice-cold saline is essential. It is also necessary for the full 'dilution' curve to be available for inspection to verify that good quality injection and detection have been achieved. Attention has also to be paid to respiration, which greatly affects flow, and four injections spaced at different points throughout the respiratory cycle are advisable. When this method is properly used it is accurate and has a reproducibility of 5–7%. It can also provide a useful check on non-invasive methods since the dilution curve lasts only 10–15 s and it can be recorded simultaneously with the measurement of cardiac output by another method.

NON-INVASIVE METHODS

None of the non-invasive methods can yet be recommended for general clinical use. They can, however, be used if their performance is properly monitored and their accuracy compared against an acceptable standard under identical conditions. Repeatability has also to be checked and it must be noted that comparison between a method which measures stroke volume with one which measures cardiac output is subject to the possibility of false correlations, since cardiac output is linearly related to heart rate.

The Doppler method is the most promising non-invasive approach available. Pulsed Doppler is acceptable but continuous wave is not. As with other methods, the primary data must be available for inspection and a good quality Doppler signal is essential. Velocity measurements from the aorta are acceptable but those which rely upon velocity across the mitral valve are not, since the area of the latter changes during inflow of blood during diastole. Analysis of the wave-form depends critically upon the quality of the fast Fourier analysis in which the acquisition time should be less than 10 ms. The method has two important limitations. First, absolute measurement of volume cannot be made since echocardiographic techniques do not allow accurate measurement of aortic diameter; the problem arises since measurement of diameter must be made at 90° to the vessel and this cannot, at present, be assured, although there is progress in this area. This limitation however does not detract from the value of the method in following stroke volume changes with time in an individual. The second limitation of the method is that it cannot yet be used during upright exercise, since movement of the heart and great vessels interferes with acquisition of the signal.

Radionuclide injections can assess cardiac output as a type of indicator-dilution method with external detection: the so-called 'first pass' method. It requires central injection of the indicator into the right atrium and the assistance of an interested physicist is essential. The gated blood pool method cannot be recommended because there are problems with resolution and detection of the cardiac borders. The decay of indicator with distance from the detectors is also a problem.

Impedance cardiography has been widely used but there are new formulae which probably invalidate earlier work. The method needs more extensive evaluation against an acceptable standard and it has to be compared with such a reliable standard method during changes in physiological conditions. Estimates of repeatability are also needed.

Echocardiography has been regularly used to estimate ventricular stroke volume. Although cavity measurements can be made with reasonable accuracy and its repeatability is satisfactory, the transformation of dimension measurements into estimates of volume rely on questionable assumptions. The correlations of their method with others shows wide variability and further work on simultaneous comparison between the echo-derived stroke volume and a standard method, such as the thermodilution technique, is required. Hence, the use of this technique to estimate stroke volume is not recommended.

Pulse-contour measurements from a peripheral artery have been used to estimate stroke volume. Measurements of the area of the pressure pulse from a peripheral artery during systole may provide an estimate of stroke volume if a factor related to arterial compliance can be obtained. Older formulae have been abandoned but newer ones have been presented which may be valid but require assessment in unanaesthetized individuals and during changes in physiological status. There is an assumption that the compliance of the great vessels does not change between calibrations and this should be checked. In any case, repeated 'calibration' by an accurate measurement of stroke volume is needed. However, the advent of non-invasive beat-by-beat estimates of blood pressure suggest that this method is worth assessing in detail.

Conclusions

Cardiac output measurement is important for clinicians. The Fick and indicator dilution methods are accurate and repeatable, but need to be used with great care. Non-invasive methods are not yet sufficiently reliable to be recommended for general use, although some can be used if they are checked against a standard method under the precise conditions in which they are to be employed in the clinic. An important responsibility rests upon the shoulders of the editors of journals to ensure that referees comment on the adequacy of the methods used for the assessment of cardiac output in papers they accept for publication.

Glossary

Aliasing
Production of an unwanted component in a signal owing to limitations in the signal generating system.

Ascending aortic blood velocity
Quantity determined on the basis of Doppler frequency shift and the angle of the ultrasound beam.

Blood velocity profile
Diversity of flow velocities as transmitted by Doppler signal from different areas of a blood vessel.

Carbon dioxide (CO_2) rebreathing
Non-invasive method for estimating cardiac output based on the Fick principle (measuring CO_2 exhalation per minute and estimating arteriovenous differences in CO_2 content).

Cardiac index (CI)
Cardiac output normalized for body surface area.

Cardiac output (CO, $\dot{Q}$)
Quantity of blood pumped by the left ventricle into the arterial system.

Cardiopulmonary or *central blood*
Quantity of blood 'contained' in the heart chambers and the pulmonary circulatory system at a given moment.

Dye dilution method (Stewart-Hamilton principle)
Determination of dye concentration curve in arterial blood drawn after 'instantaneous' central injection of indicator dye (e.g. indocyanine green); the area under the curve are indicative of blood flow through the area under study and hence of cardiac output.

Ejection fraction (EF)
Stroke volume as a percentage of left ventricular diastolic volume (ratio stroke volume: end-diastolic volume)

Echocardiography
M-mode: Estimation of cardiac dimensions including different structures during the cardiac cycle by ultrasound reflection; stroke volume may be derived by approximating the difference between diastolic and systolic volumes of the left ventricle.

Doppler-flow: Estimation of stroke volume by Doppler frequency shift.

Electromechanical systole (QS_2)
Interval between onset of the electrocardiographic QRS complex and closure of the aortic valve.

Fick principle
Blood flow through an organ (e.g. lungs) is equal to the amount of a substance (e.g. oxygen) absorbed by the blood flowing through the organ divided by the difference in concentration of the substance (oxygen) between the blood entering and leaving the organ.

Impedance
The complex of forces counteracting forward blood flow by ventricular ejection.

Impedance cardiography
Method for assessment of stroke volume, based on measurement of changes in transthoracic electrical impedance to a weak high frequency signal, during the cardiac cycle.

Insonation angle
Directional angle of ultrasonic (Doppler) beam to the longitudinal axis of a blood vessel.

Left ventricular ejection time (LVET)
Interval between opening and closure of the aortic valve.

Left ventricular mass (LVM)
Echocardiographic estimation of left ventricular muscle weight by cubing mono- or bidimensional measurements of wall thickness and assuming a valve for the specific gravity of the muscle.

Magnetic resonance imaging (MRI)
Imaging technique based on intrinsic differential proton radiofrequency between moving blood and surrounding structures.

Mixed venous blood
Venous blood drawn from a site where blood arriving from areas with different oxygen consumption has become homogeneous with respect to oxygen saturation (e.g. right ventricle or pulmonary artery).

Oxygen uptake
Amount of ambient oxygen from the environment taken up by the lungs over a given period of time.

Position emission tomography (PET)
Imaging technique based on the use of positron emitters (^{82}Rb, ^{18}F etc.) allowing calculation or estimation of ventricular chamber size, wall thickness and stroke volume, as well as the characterization of certain metabolic features.

Pre-ejection period (PEP)
Isometric ventricular contraction as derived from the interval between the onset of ventricular depolarization and the opening of the aortic valve.

Pulse contour assessment (of stroke volume)
Method relating flow to pressure wave using impedance characteristics of the large vessels.

Steady state
Physiological equilibrium attained after emotional and/or physical perturbations (such as those induced by the initiation of investigational procedures) have receded.

Stewart-Hamilton principle
(See dye dilution).

Stroke distance
Distance covered by the movement of a column of blood through the ascending aorta during one cardiac cycle; equals stroke volume when divided by aortic cross-sectional area.

Stroke volume (SV)
Amount of blood ejected per heart beat.

Stroke volume index (SVI)
Stroke volume normalized for body surface area.

Systolic time intervals
Components of systole as derived from simultaneous recordings of ECG, phonocardiogram, carotid pulse tracings and/or echocardiogram with the aim of assessing left ventricular performance.

Thermodilution technique
Method based on the Stewart-Hamilton principle using the introduction of a thermistor into the pulmonary artery and the registration of the temperature curve produced by 'instantaneous' injection of a precise quantity of ice-cold saline or glucose into the right atrium.

Total peripheral vascular resistance (TPR)
Resistance to flow in the arterial system (outside the lungs) as estimated by dividing mean arterial pressure by cardiac output.

Total peripheral vascular resistance index (TPRI)
Total peripheral vascular resistance normalized for body surface area.

Ultrafast computed tomography (cine-CT)
Scanning technique allowing total mapping of left ventricular size changes within one cardiac cycle after contrast injection.

Index

Printed in Great Britain by Henry Ling Ltd., at the Dorset Press, Dorchester, Dorset